Progress in Theoretical Computer Science

Christian Prehofer

Solving Higher-Order Equations

From Logic to Programming

Birkhäuser
Boston • Basel • Berlin

Christian Prehofer
Institut für Informatik, SB3
München, Germany 80290

Library of Congress Cataloging-in-Publication Data

Prehofer, Christian, 1967-
Solving higher-order equations : from logic to programming / Christian Prehofer.
p. cm. -- (Progress in theoretical computer science)
Includes bibliographical references and index.
ISBN 0-8176-4032-0. -- ISBN 3-7643-4032-0
1. Computer logic. 2. Declarative programming. 3. Logic, Symbolic and mathematical. I. Title. II. Series.
QA76.9.L63P74 1997
005.13'1--dc21 97-31142
CIP

AMS Subject Classifications: 0B3-XX, 03C-XX, 03D-XX
ACM Subject Classifications: D.1-3., E.2., F.2-4., I.1-2.

Printed on acid-free paper

Birkhäuser

ISBN 0-8176-4032-0
ISBN 3-7643-4032-0
Typeset by the author in LaTeX.
Printed and bound by Quinn-Woodbine, Woodbine, NJ.
Printed in the U.S.A.

9 8 7 6 5 4 3 2 1

Contents

1 Introduction **1**

2 Preview **7**

2.1 Term Rewriting 7
2.2 Narrowing 8
2.3 Narrowing and Logic Programming 10
2.4 λ-Calculus and Higher-Order Logic 11
2.5 Higher-Order Term Rewriting 13
2.6 Higher-Order Unification 14
2.7 Decidability of Higher-Order Unification 16
2.8 Narrowing: The Higher-Order Case 17
2.8.1 Functional-Logic Programming 20
2.8.2 Conditional Narrowing 21

3 Preliminaries **25**

3.1 Abstract Reductions and Termination Orderings 25
3.2 Higher-Order Types and Terms 27
3.3 Positions in λ-Terms 30
3.4 Substitutions 32
3.5 Unification Theory 33
3.6 Higher-Order Patterns 34

4 Higher-Order Equational Reasoning **37**

4.1 Higher-Order Unification by Transformation 37
4.2 Unification of Higher-Order Patterns 43
4.3 Higher-Order Term Rewriting 46
4.3.1 Equational Logic 49
4.3.2 Confluence 50
4.3.3 Termination 52

5 Decidability of Higher-Order Unification 55
5.1 Elimination Problems 55
5.2 Unification of Second-Order with Linear Terms 62
5.2.1 Unifying Linear Patterns with Second-Order Terms . 62
5.2.2 Extensions 65
5.3 Relaxing the Linearity Restrictions 70
5.3.1 Extending Patterns by Linear Second-Order Terms . 70
5.3.2 Repeated Second-Order Variables 72
5.4 Applications and Open Problems 75
5.4.1 Open Problems 77

6 Higher-Order Lazy Narrowing 79
6.1 Lazy Narrowing 81
6.2 Lazy Narrowing with Terminating Rules 87
6.2.1 Avoiding Lazy Narrowing at Variables 88
6.2.2 Lazy Narrowing with Simplification 90
6.2.3 Deterministic Eager Variable Elimination 93
6.2.4 Avoiding Reducible Substitutions by Constraints . . 94
6.3 Lazy Narrowing with Left-Linear Rules 96
6.3.1 An Invariant for Goal Systems: Simple Systems . . . 97
6.3.2 A Strategy for Call-by-Need Narrowing 103
6.3.3 An Implementational Model 110
6.4 Narrowing with Normal Conditional Rules 111
6.4.1 Conditional Rewriting 112
6.4.2 Conditional Lazy Narrowing with Terminating Rules 114
6.4.3 Conditional Lazy Narrowing with Left-Linear Rules 118
6.5 Scope and Completeness of Narrowing 118
6.5.1 Oriented versus Unoriented Goals 120

7 Variations of Higher-Order Narrowing 121
7.1 A General Notion of Higher-Order Narrowing 122
7.2 Narrowing on Patterns with Pattern Rules 125
7.3 Narrowing Beyond Patterns 128
7.4 Narrowing on Patterns with Constraints 130

8 Applications of Higher-Order Narrowing 135
8.1 Functional-Logic Programming 136
8.1.1 Hardware Synthesis 137
8.1.2 Symbolic Computation: Differentiation 138
8.1.3 A Functional-Logic Parser 141
8.1.4 A Simple Encryption Problem 143

8.1.5 "Infinite" Data-Structures and Eager Evaluation . . . 144
8.1.6 Functional Difference Lists 145
8.1.7 The Alternating Bit Protocol 146
8.2 Equational Reasoning by Narrowing 149
8.2.1 Program Transformation 149
8.2.2 Higher-Order Abstract Syntax: Type Inference . . . 151

9 Concluding Remarks **153**
9.1 Related Work 154
9.1.1 First-Order Narrowing 154
9.1.2 Other Work on Higher-Order Narrowing 155
9.1.3 Functional-Logic Programming 157
9.1.4 Functional Programming 158
9.1.5 Higher-Order Logic Programming 158
9.2 Further Work 159
9.2.1 Implementation Issues 159
9.2.2 Other Extensions 160

Bibliography **163**

Index **182**

To Andrea

The Emperor counsels simplicity.
First principles.
Of each particular thing, ask:
What is it in itself,
in its own constitution?
What is its causal nature?

Dr. Hannibal Lecter, in *The Silence of the Lambs*,
Thomas Harris

List of Figures

1.1 Declarative Programming Paradigms 2

2.1 Decidability of Higher-Order Unification 15
2.2 A Framework for Higher-Order Narrowing 19

3.1 $\lambda x_1, x_2.F(a,b)$ as Binary Tree 31
3.2 $\lambda x_1, x_2.F(a,b)$ as n-nary Tree 31

4.1 System PT for Higher-Order Pre-Unification 38
4.2 Search Tree with System PT 41
4.3 System PU for Pattern Unification 45
4.4 Equational Theory of a GHRS R 50

5.1 Results on Second-Order Unification 56
5.2 System EL for Eliminating Bound Variables 58

6.1 Dependencies of Lazy Narrowing Refinements 80
6.2 System LN for Lazy Narrowing 83
6.3 Deterministic Constructor Rules 87
6.4 Rules for Lazy Narrowing with Constraints (LNC) 95
6.5 Rules for Conditional Lazy Narrowing (CLN) 115

7.1 System NC for Narrowing with Constraints 131

8.1 Rules for Hardware Synthesis 138
8.2 Rules R_d for Symbolic Differentiation 139
8.3 Communication Model for Authorization 143
8.4 Alternating Bit Protocol 147
8.5 Rules for the Alternating Bit Protocol 148

Chapter 1

Introduction

This monograph develops techniques for equational reasoning in higher-order logic. Due to its expressiveness, higher-order logic is used for specification and verification of hardware, software, and mathematics. In these applications, higher-order logic provides the necessary level of abstraction for concise and natural formulations. The main assets of higher-order logic are quantification over functions or predicates and its abstraction mechanism. These allow one to represent quantification in formulas and other variable-binding constructs.

In this book, we focus on equational logic as a fundamental and natural concept in computer science and mathematics. We present calculi for equational reasoning modulo higher-order equations presented as rewrite rules. This is followed by a systematic development from general equational reasoning towards effective calculi for declarative programming in higher-order logic and λ-calculus. This aims at integrating and generalizing declarative programming models such as functional and logic programming. In these two prominent declarative computation models we can view a program as a logical theory and a computation as a deduction.

Equational Logic and Term Rewriting

Rewrite systems are directed equations and allow for simplification of terms or expressions. Rewriting offers a simple operational semantics for programming applications and furthermore enables effective reasoning methods. For instance, term rewriting provides a basis for functional programming languages, e.g. Haskell and SML. Also, goal solving in logic programming can be viewed as equational reasoning, which is the common starting point for the integration of functional and logic programming. Furthermore, rewriting allows one to study evaluation strategies and to reason about programs. For

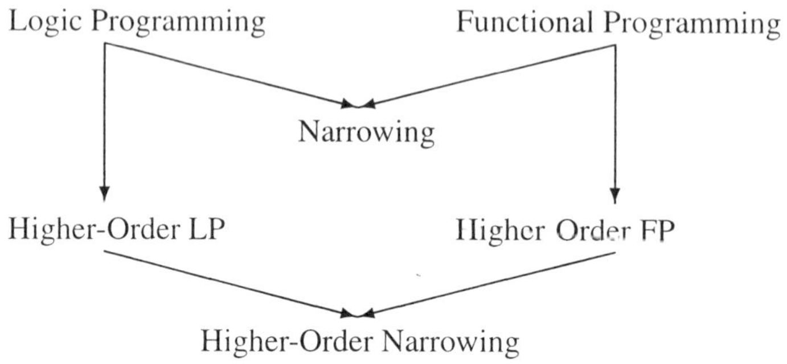

Figure 1.1: Declarative Programming Paradigms

instance, confluent and terminating rewrite rules guarantee that every term has a unique normal form. This is often used to give meaning to programs, but it is also valuable for equational reasoning.

Higher-order rewriting, in particular, is very suitable for symbolic computation with complex structures in mathematics and programming. For instance, the kernel of Mathematica^{TM} is a particular version of higher-order rewriting. In the setting of higher-order rewriting, we develop higher-order narrowing, a technique for solving goals with existentially quantified variables. Based on narrowing, our methods for higher-order equational reasoning generalize and advance first-order techniques. The focus on equational logic and term rewriting allows the development of effective theorem proving techniques that are not possible in full higher-order logic.

Integrating Functional and Logic Programming

Building on equational logic, we pursue the integration of declarative programming paradigms in higher-order logic. Figure 1.1 shows the relations between existing programming paradigms with respect to higher-order aspects. Whereas higher-order programming is standard in functional programming, logic programming is in large part still tied to the first-order world. Only a few higher-order logic programming languages, most notably λ-Prolog, use higher-order logic for logic programming and have shown its practical utility. For the integration of both declarative paradigms, many calculi based on first-order narrowing have been developed (for a survey see [Han94b]). Whereas logic programming is based on predicate logic and functional programming on λ-calculus, their integration usually relies on equational logic and term rewriting.

This integration of declarative programming paradigms has been examined extensively and has led to several implementations, but the foundations for the higher-order case have been neglected. Some currently developed languages stress functional evaluation over equational reasoning and completeness in a higher-order setting. Their incompleteness means that existing solutions for some goals cannot be computed.

We use higher-order narrowing as an inference rule which provides for a nice generalization of the inference mechanisms in both logic and functional programming. As narrowing solves equations with existentially quantified or "logic" variables, this can be viewed as extending a functional programming language by logic variables. If no logic variables are present, narrowing simply coincides with functional evaluation. Note that our setting also captures the main concepts of higher-order logic programming, where λ-terms are used as data structures as done here.

Decidable Higher-Order Unification

One of the major obstacles for reasoning in higher-order logic is that unification is undecidable in the higher-order case. Unification serves as the basic inference engine in logic programming and theorem provers and is also used for narrowing. On the practical side, higher-order logic programming pioneered in the use of higher-order unification for logic programming and has shown its practical utility despite its undecidability. However, several works [Smo86, Loc93, SJ92] on functional-logic programming claim that higher-order unification cannot be used due to its undecidability. Hence we examine higher-order unification and establish decidable cases that are sufficient for programming applications. More precisely, these results permit second-order functional-logic programming where common unification algorithms terminate.

A Framework for Higher-Order Narrowing

In the main part of this work we develop the framework for higher-order narrowing. Starting with completeness results for general equational reasoning, we systematically refine the narrowing calculi towards programming applications. These refinements reduce the degree of non-deterministic search by using particular properties of rewrite rules. This stepwise development follows two directions. On the one hand, we pursue general equational reasoning with terminating rules, and on the other hand functional-logic programming with left-linear rewrite rules. For instance, left-linearity of rewrite rules in functional programs leads to decidable unification in the second-order case and to a new strategy for needed narrowing, where intermediate goals are only solved when needed. The refinements culminate in a call-by-need narrowing strategy, generalizing the functional evaluation model to this context.

The main steps are summarized in the following:

- Decidable second-order unification problems which can be applied to functional-logic programming.
- Narrowing calculi which are suitable in the higher-order case.
- Several refinements of lazy narrowing for general equational reasoning, e.g. deterministic simplification, conditional narrowing.
- A new approach to (higher-order) functional-logic programming based on narrowing, for which we establish a call-by-need narrowing strategy.

This framework is furthermore extended to conditional rewrite rules, which are useful in many applications. We show that a simple class of conditional rewrite rules suffices in our higher-order setting, as some common programming constructions are encoded in higher-order functional style instead of logic programming techniques.

Our development of equational reasoning by narrowing is complemented by a chapter discussing alternative approaches to narrowing. We show that other common techniques for equational reasoning are not suitable in the higher-order case.

Applications

We demonstrate the application of higher-order narrowing by a set of examples ranging from symbolic differentiation and program transformation to programming techniques in λ-calculus. In particular, we discuss some novel programming concepts in higher-order functional and logic programming. For instance, quantified goals are possible and functional objects, e.g. programs or hardware circuits, can be synthesized. Furthermore, some examples, such as the simple parser and the alternating bit protocol show the utility of having both paradigms, functional and logic programming, in one setting.

How to Read This Book

The structure of the book is as follows. The next chapter gives an informal outline of this work. More detailed overviews of the results can be found at the beginning of each chapter. Chapter 3 presents simply typed λ-calculus and other basic preliminaries. An introduction to higher-order unification and term rewriting follows in Chapter 4. Chapter 5 develops decidable classes of second-order unification.

Higher-order lazy narrowing in Chapter 6 is the main subject of this work and can be read, with a few exceptions, independently of Chapter 5. This is followed by a discussion on alternative approaches to higher-order narrowing in Chapter 7. Examples for higher-order narrowing are shown in Chapter 8 and Chapter 9 concludes with discussions and comparisons.

Many of the results here have been presented in earlier papers. The first results on a framework for higher-order narrowing can be found in [Pre94b], which have been developed further for equational reasoning in [Pre95b] and for functional logic programming in [Pre95a]. The decidability results for higher-order unification were presented in [Pre94a]. The present book is an extended presentation of the author's doctoral thesis [Pre95c]. A compact overview of some results can be found in [Pre97].

Acknowledgments

I wish to thank Tobias Nipkow for his continuous support. I am grateful to many friends and colleagues for comments and discussions on the subject. They include Hitoshi Ohsaki, Aart Middeldorp, Olaf Müller, Joachim Niehren, Jaco van de Pol, Mario Rodríguez-Artalejo, Heinrich Hußmann, Oscar Slotosch, Cornel Klein, Konrad Slind, Max Moser, Andreas Werner, Michael Hanus, Gilles Dowek, Vincent van Oostrom, Michael Kohlhase, and Gérard Huet. Furthermore, I thank Robert Furtner for his efforts on implementing parts of this work. I am indebted to the anonymous reviewers for their careful and detailed feedback.

Manfred Broy and Tobias Nipkow provided the fruitful environment that made this work possible. This includes many others of the Munich research group as well. Finally, I want to thank my relatives and friends, particularly Andrea, for enduring me on this adventure.

Chapter 2

Preview

In this chapter, we informally introduce the main concepts and outline the contributions of this work. Precise definitions are presented in later chapters. We proceed from first-order term rewriting and narrowing to higher-order unification and higher-order narrowing.

2.1 Term Rewriting

Term rewriting is a model of computation. Rewriting is based on the idea of "replacing equals by equals". For this purpose, equations between terms are oriented into rewrite rules. For instance, the equations $0+Z=Z$ and $X+succ(Y)=succ(X+Y)$ form a specification of the function $+$, assuming the term constructors 0 and *succ*. We orient the equations into the following two rules:

$$R = \left\{ \begin{array}{rcl} 0+Z & \rightarrow & Z \\ X+succ(Y) & \rightarrow & succ(X+Y) \end{array} \right\}$$

With orientation, we gain an operational model: reduction. We can reduce a term with the rules of R, e.g.:

$$a+succ(0+b) \longrightarrow a+succ(b) \longrightarrow succ(a+b)$$

where a and b are some constants. These two reduction steps reduce $a+succ(0+b)$ to its normal form $succ(a+b)$ w.r.t. R. Notice that this is an abstract model of computation; it is not directly used as a programming language since reduction is in general not deterministic. Programming languages usually restrict rewrite systems to be confluent, which implies that normal forms are unique, e.g. $succ(a+b)$ above. Then it suffices to follow only a particular reduction strategy.

Term rewriting is, for instance, useful as an abstract model for symbolic reasoning with equations and for the analysis of programming languages. In particular, term rewriting is an abstract model of first-order functional programming languages. Current languages such as LISP variants [Ste90], Haskell [PHA$^+$96] or SML [MTH90] are higher-order and originate from the λ calculus.

A very common and useful extension of rewriting is to add conditions to the rules. For instance, consider the rules

$$\begin{array}{lcll} fib(X) & \to & fib(X-1)+fib(X-2) & \Leftarrow X > 1 \\ fib(X) & \to & 1 & \Leftarrow X \leq 1 \end{array}$$

for computing the Fibonacci numbers. In such a setting, the condition has to be established (again by rewriting), before a rewrite step can be applied.

The simplicity of the concept of term rewriting has attracted a lot of research concerning properties of rewrite systems such as termination or confluence. For surveys we refer to [DJ90, Klo92, BN]. Apart from programming, well developed applications are theorem proving, including both automatic systems [Hsi85] and interactive systems [Gor88, Pau94], program synthesis via completion [Bac91] and algebraic specifications [GTW89, EM85, FH91, Wir90].

2.2 Narrowing

Starting from an equational specification, it is often not only desirable to evaluate terms, but also to solve equations. For instance, with the rules of R, a simple goal is to ask for what values of X the equation

$$succ(X+a) =^? succ(a)$$

holds. *Narrowing* is a general mechanism for solving such goals in a systematic way. The goal is to find values for X such that the equation holds w.r.t. the equational theory of R. This is also called equational or R-unification.

For this purpose, we use equational goals of the form

$$s \to^? t$$

where a substitution θ is a solution if

$$\theta s \xrightarrow{*} t.$$

We assume for initial goals that t is ground, e.g. the constant *true*, but during a computation the right-hand side may have free variables, as shown later. For equational reasoning with undirected goals of the form $s =^? t$, we can just add a rule $X =^? X \to true$ and solve $s =^? t \to^? true$, where $=^?$ is a new symbol. Note that this only works for confluent systems. The focus on directed goals simplifies many technicalities and is essential for programming applications as shown later.

Whereas term rewriting searches for matches of a rule, narrowing uses *unification* to find an instance of a term such that a rewrite step applies. For instance, unifying the left-hand side of the first rule of R, $0+Z$, with $X+a$ yields a solution by the substitution

$$\theta = \{X \mapsto 0, Z \mapsto a\}.$$

Then we have the narrowing step

$$succ(X+a) \leadsto_{\theta}^{0+Z\to Z} succ(a). \tag{2.1}$$

The gist of narrowing is that it need not be applied to variable subterms, which would be highly non-deterministic. For narrowing the restriction to R-normalized solutions, which map variables to terms in R-normal form, implies that this is not needed. Yet this is just the starting point for further optimizations.

Compared to paramodulation [RW69, Bra75], an early precursor for equational reasoning, narrowing [Sla74] assumes rewrite rules instead of undirected equations. Research on first-order narrowing was initiated by the papers of Fay [Fay79] and Hullot [Hul80]. Hullot first showed correctness and completeness of narrowing, which roughly states that if a goal $s \to^? t$ has a solution θ, i.e. $\theta s \xrightarrow{*}^{R} t$, then there is a sequence of narrowing steps such that the computed solution is equal to or more general than θ. With the kind of goals we consider, this result only deals with equational matching, but unification is easy to encode, as shown above (see also Section 6.5). Hence narrowing serves as a complete method for unification modulo a theory given by a convergent term rewriting system: narrowing is complete w.r.t. normalized substitutions and for every substitution there exists an equivalent normalized one.

Narrowing forms the underlying computation rule for functional-logic programming languages [Red85, DO90]. For instance, logic programming can be viewed as narrowing [BGM88] and work on integrating logic and functional programming is usually based on narrowing. Many of the early proposals for functional-logic programming can be found in [DL86]. When using narrowing as a programming language model, reduction is viewed as evaluation.

As the search space of naive narrowing is very large, there exists an abundance of refinements which remove redundant narrowing derivations (see [Han94b, MH94] for overviews). For convergent rewrite systems, there is a strategy [BKW93] that is optimal in the sense that no solution is computed twice. For a restricted class of term rewriting systems, which suffices for simple programming languages, there exists a strategy [AEH94] that computes reductions of minimal length.

Apart from the notion of narrowing explained above, there exists another notion of narrowing, called *lazy narrowing*. To avoid confusion, we call the first notion *plain narrowing*. Plain narrowing searches for an instance such that some subterm can be rewritten. In contrast, lazy narrowing integrates the rules of unification into narrowing. The idea is to simplify terms by unification until only rewrite steps at the outermost position have to be considered in a "lazy" fashion.

For instance, to model the (plain) narrowing step in (2.1) by lazy narrowing, we start with a goal

$$succ(X+a) \rightarrow^? succ(a)$$

and look for a solution θ such that $succ(\theta X + a)$ rewrites to $succ(a)$. We first apply a decomposition step on the constructor symbol *succ*, yielding the subgoal

$$X+a \rightarrow^? a.$$

Then a lazy narrowing step applies at the function symbol $+$ with the rule $0+Z \rightarrow Z$. The unification of the subterms of the rewrite rule with the goal is delayed by generating two new goals for the unification of $X+a$ with $0+Z$:

$$X \rightarrow^? 0, a \rightarrow^? Z, Z \rightarrow^? a$$

Lazy narrowing employs such steps only at the root position of a term. In general, the newly added subgoals must again be solved modulo R. In this example, it suffices to take the direct syntactic solution, i.e. $\{X \mapsto 0, Z \mapsto a\}$.

Most papers on narrowing and functional-logic languages employ variations of these two notions of narrowing. Plain narrowing is mostly used for terminating rewrite systems with equational semantics [Han91]. Alternatively, narrowing is also used with denotational semantics [Red85], based on strict equality: two terms are equal if they evaluate to the same constructor or data term. For this semantics, there exist completeness results for narrowing with non-terminating rules, see for instance [MNRA92, GHR92].

2.3 Narrowing and Logic Programming

The relationship between logic programming [CM84, Llo87] and narrowing is well examined. Most approaches to functional-logic programming are based on narrowing and aim at extending logic programming by functions [Han94b]. In such languages, narrowing replaces resolution as the basic mechanism of inference. The idea is simple: view predicates as functions and horn clauses as rules with conditions. That is, a clause

$$P :- Q_1, Q_2, \ldots, Q_n$$

is written equivalently as

$$P \to true \Leftarrow Q_1 \to true, Q_2 \to true, \ldots, Q_n \to true.$$

It has been shown that narrowing, with conditional or unconditional equations, can simulate logic programming and vice versa [BGM88, Huß93]. There exist however more advanced refinements for narrowing that utilize the determinism of functional programs to a large extent. For instance, functional-logic programming with normalization is more effective than pure logic programming, see e.g. [CF91, Han92]. These refinements use deterministic evaluation, which is possible for convergent rewrite rules.

In pure logic programming, functions are encoded as predicates. The functional version is often more concise, as functions can be nested in contrast to predicates. For instance, consider the clause

$$fib_P(s(s(X)), YZ) :- fib_P(s(X), Y), fib_P(X, Z), plus_P(Y, Z, YZ),$$

where the predicate $plus_P(Y, Z, YZ)$ holds if $YZ = Y + Z$. This becomes

$$fib(s(s(X)) = fib(s(X)) + fib(X)$$

in functional-logic programming. Notice also that logic programming needs additional local variables.

2.4 λ-Calculus and Higher-Order Logic

Before introducing higher-order term rewriting, we take a brief tour of the basic concepts of λ-calculus [HS86, Bar84]. Whereas first-order terms in first-order logic describe (concrete) functions (or predicates) on data, higher-order logic also allows one to reason about functional objects (or predicates). Thus functions may take functions as arguments and also return functions. In programming languages, this is often described as "functions as first-class citizens". Since λ-calculus nicely supports many basic concepts of programming languages, it is used for their formalization and serves as an abstract computation model. Originally, (untyped) λ-calculus was introduced [Chu40] as a theory of functions and serves as a general model of computable functions. Here, we use simply typed λ-calculus as the underlying data structure and add rewrite rules, similar to first-order rewriting. Since reductions of simply typed λ-calculus do terminate, it is not a full computational model but is well suited as our underlying term structure.

Coming from first-order terms, λ-calculus adds λ-abstractions, which construct new functions. An abstraction $\lambda x.t$ is hence a new function which abstracts over the variables x in t. Such abstractions are often viewed as "nameless" functions. In $\lambda x.t$, λx is called the binder and x is called a bound variable. To illustrate λ-abstractions consider a function definition found in programming languages:

$$\texttt{function}\ plus\ x\ y = x + y$$

This definition can be written equally as

$$\texttt{function}\ plus = \lambda x.\lambda y.x + y,$$

where the right-hand side is a functional term.

Due to explicit abstractions, one has to generalize application as in first-order logic. In λ-calculus, any term t of type σ can be applied to a functional term a of type $\sigma \to \sigma'$ written as $s\ t$. Parameter passing in λ-calculus is performed by the rule of β-reduction:

$$(\lambda x.t)t \succ_\beta \{x \mapsto s\}t$$

Note that partial applications are also allowed, e.g. *plus* 1 is a function that increments its argument by one. If possible, we still write terms in first-order fashion, e.g. $(f\ 1)\ 2$ is written as $f(1,2)$.

Another important contribution of λ-calculus is to view the bound variables in abstractions as nameless parameters. Hence we consider two terms equal if they are equal modulo renaming of bound variables. This is achieved by the rule of α-conversion. For instance, $\lambda x.f(x) =_\alpha \lambda y.f(y)$. An important consequence is that substitution cannot interact with bound variables. For instance, if we replace Y by $s(x)$ in a term $\lambda x.g(Y)$, then $\lambda x.g(Y)$ must (implicitly) be renamed, otherwise x in $x(s)$ would be captured by λx. Furthermore, we work with extensional λ-calculus, where a functional term t is identified with the function $\lambda x.t(x)$, called η-conversion, if x is not free in t. Assuming η-conversion, we can also show the principle of extensionality. This means that two functions are considered equal, if they behave equally for all arguments.

When computing with λ-terms, the underlying logic is higher-order and allows quantification over variables of functional type (see [And86] for a comprehensive treatment). As we focus on equational logic, we do not go into the details of higher-order logic. Thus we only consider (quantified) equations between higher-order or λ-terms including higher-order variables. A property which we will often use is that universally quantified equations can be modeled by equality between functions. For instance, $\forall x.f(x) = g(x)$ is logically equivalent to $\lambda x.f(x) = \lambda x.g(x)$.

2.5 Higher-Order Term Rewriting

Higher-order term rewriting is the natural extension of first-order rewriting to reasoning with higher-order equations. Starting with the work of Klop [Klo80], there exist several notions of higher-order term rewriting [Nip91a, Oos94, Raa93]. This interest in higher-order rewriting follows the progress in its applications, for instance functional languages and theorem provers. In this work, we follow the approach in [Nip91a]: we consider simply-typed λ-terms in β-normal form and view the reductions of λ-calculus as implicit operations. For instance, immediate substitution of arguments via β-reduction, e.g. $(\lambda x.f(x))a =_\beta f(a)$, is generally assumed. Furthermore, we compute modulo α-conversion, i.e. the renaming of bound variables.

The expressiveness of higher-order term rewriting easily deals with scoping constructs. We can for instance handle quantifiers as data objects; consider e.g. pushing quantifiers inside boolean expressions:

$$\forall x.P \wedge Q(x) \to P \wedge \forall x.Q(x)$$

In this example a quantifier $\forall$ is a constant of type $(term \to bool) \to bool$, where $\forall(\lambda x.P)$ is written as $\forall x.P$ for brevity. Notice that the variable conventions of λ-calculus allow for a concise statement of the first rule: the variable P in $\lambda x.P \wedge Q(x)$ represents a term not containing the bound variable x. If a term is substituted for x by β-reduction, it is appropriately renamed.

As another example for the utility of higher-order programming, consider symbolic differentiation. The function $diff(F,X)$, as defined below, computes the differential of a function F at a point X.

$$\begin{array}{rcl}
diff(\lambda y.F,X) & \to & 0 \\
diff(\lambda y.y,X) & \to & 1 \\
diff(\lambda y.sin(F(y)),X) & \to & cos(F(X)) * diff(\lambda y.F(y),X) \\
diff(\lambda y.F(y)+G(y),X) & \to & diff(\lambda y.F(y),X) + diff(\lambda y.G(y),X) \\
diff(\lambda y.F(y)*G(y),X) & \to & diff(\lambda y.F(y),X) * G(X) + \\
 & & diff(\lambda y.G(y),X) * F(X)
\end{array}$$

$$diff(\lambda y.ln(F(y)),X) \rightarrow diff(\lambda y.F(y),X)/F(X)$$

With these rules, we can for instance compute:

$$\begin{array}{ll} diff(\lambda y.sin(sin(y)),X) & \longrightarrow \\ cos(sin(X)) * diff(\lambda y.sin(y),X) & \longrightarrow \\ cos(sin(X)) * cos(X) * diff(\lambda y.y,X) & \longrightarrow \\ cos(sin(X)) * cos(X) * 1 & \end{array}$$

In contrast, first-order term rewriting only permits a limited, first-order version of *diff*, as e.g. in [Bac91, SS86] (see Section 8.1.2). For instance, the first rule cannot be expressed directly and must be coded by some means. We will use this example throughout Chapter 6 and come back to it in Chapter 8.

Apart from such high-level computations, an important application of higher-order rewriting is to model the basic mechanisms of current, higher-order functional programming languages such as SML or Haskell. Clearly, higher-order rewriting is more expressive, since it permits explicit abstractions in the left-hand sides (as in the above example), which is not possible in functional programming.

In recent years many results for first-order term rewriting have been lifted to the higher-order case. Among the results obtained for higher-order rewriting are a critical pair lemma for higher-order term rewriting systems (HRS), confluence of orthogonal HRS, and termination criteria.

2.6 Higher-Order Unification

For the step from first-order to higher-order narrowing, we examine another important ingredient: higher-order unification. Higher-order unification is a powerful method for solving equations between higher-order λ-terms modulo the conversions of λ-calculus. In particular, bound variables must be treated correctly: the unification problem

$$\lambda x.sin(F(x)) =^? \lambda x.sin(cos(x))$$

has solution $\{F \mapsto \lambda y.cos(y)\}$, whereas

$$\lambda x.F =^? \lambda x.sin(cos(x))$$

is unsolvable, as the left-hand side does not depend on x. Another difference to the first-order case is that unification may produce several incomparable unifiers, called maximally general unifiers. For instance, the unification problem

$$F(a) =^? a$$

Order	Unification Problem			
	Unification	Patterns	Monadic	Matching
1	decidable			
2	undecidable Goldfarb '81 Farmer '91	⋮	decidable Farmer '88	decidable Huet '73
3	undecidable Huet '73 Lucchesi '72	⋮	undecidable Narendran '90	decidable G. Dowek '92
∞	⋮	decidable D. Miller '91	⋮	?

Figure 2.1: Decidability of Higher-Order Unification

has solutions $\{F \mapsto \lambda x.x\}$ and $\{F \mapsto \lambda x.a\}$. In general, there may even be an infinite set of solutions. Consider the problem

$$F(f(a)) =^? f(F(a))$$

which has the solutions

$$\{F \mapsto \lambda x.f^n(a)\},\ n \geq 0,$$

where $f^0(X) = X$ and $f^{n+1}(X) = f(f^n(X))$.

With the algorithm for higher-order unification, goals are solved by decomposition to smaller goals and by computing partial solutions for free variables. The latter is more involved in the higher-order case, as bound variables have to be considered.

Higher-order unification is currently used in theorem provers like Isabelle [Pau90], TPS [AINP90], Nuprl[1] [CAB$^+$86] and for higher-order logic programming in the language λ-Prolog [NM88]. Other applications of higher-order unification include program synthesis [Hag91b] and machine learning [Har90, DW88, Hag91a].

The first complete set of rules for higher-order unification was presented by Jensen and Pietrzykowski [Pie73, JP76]. The undecidability of higher-order unification was first shown by Huet [Hue73] and Lucchesi [Luc72]. It took several years until the undecidability was shown for the second-order case by Goldfarb [Gol81]. Farmer [Far91] refined this result by showing that only one symbol of arity two is needed and by giving a bound on the

[1] Nuprl uses only second-order pattern matching.

number of variables needed to express an undecidable problem by second-order unification.

Figure 2.1 presents an overview of known decidability results for higher-order unification. The column labeled monadic refers to the unification of terms with unary function symbols only. Monadic second-order unification is decidable [Far88]. This problem can in fact be related to unification modulo associativity, which was shown to be decidable by Makanin [Mak77]. Again, the third-order monadic case is undecidable [Nar89].

Huet [Hue75] already conjectured that higher-order matching, i.e. unification with a term containing no free variables, is decidable, but the problem is still open. Some progress has been made by Dowek [Dow92], who showed the decidability of third-order matching. Furthermore, fourth-order matching is shown to be decidable by Padovani [Pad95]. Wolfram [Wol93] presents a terminating algorithm for higher-order matching, but was not able to show its completeness.

Dale Miller, as indicated in the column labeled patterns, discovered a class of λ-terms, called higher-order patterns, with decidable and even unitary unification, i.e. if some unifier exists then there exists a most general unifier. A term is a higher-order pattern if each free variable has distinct bound variables as arguments. For instance, $\lambda x, y.F(x, y)$ is a higher-order pattern, but $\lambda x.G(H(x))$ is not.

Patterns behave like first-order terms in many respects, e.g. unification is not only unitary but also of linear complexity [Qia93]. The properties of patterns will be very important for developing our narrowing calculi later, as several common first-order arguments still work for patterns, but not beyond.

Full higher-order unification is highly intractable: there do not exist maximally general unifiers. In other words, there are infinite chains of unifiers, one more general than the other. This is called nullary unification. As noted by Huet [Hue75], this was first observed by Gould [Gou66]. The idea of pre-unification by Huet [Hue75] was a major step towards practically usable systems: pre-unification delays a particular class of equations that is known to be solvable and permits the enumeration of a complete set of unifiers without any redundancy. This is important for any practical application. Pre-unification is still infinitary, i.e. there may be an infinite set of unifiers for two terms.

2.7 Decidability of Higher-Order Unification

Since higher-order unification is undecidable in general, we are interested in classes where higher-order unification is decidable. In many works concerning higher-order unification [Wol93, BS94, Nad87, Pau94], it is observed that non-termination of higher-order (pre-)unification occurs very rarely in practice.

The main restriction we impose is linearity, i.e. we require that some variables may not occur repeatedly. We show that the unification of a linear higher-order pattern s with an arbitrary second-order term that shares no variables with s is decidable and finitary. In particular, we do not have to resort to pre-unification, as equations with variables as outermost symbols on both sides (flex-flex pairs) can be finitely solved in this case. A few extensions of this unification problem are still decidable.

The main application of this result is the unification of linear left-hand sides of rewrite rules with second-order terms, as employed in higher-order narrowing. Note that functional programs have linear left-hand sides. For instance, consider the function

$$map(F, [X|Y]) = [F(X)|map(F, Y)].$$

Since it has the non-pattern $F(X)$ on the right-hand side, rewriting with this rule may yield non-pattern terms. Thus higher-order unification is needed for the unification with a left-hand side of a rewrite rule. So far, most functional logic languages even with higher-order terms only use first-order unification, e.g. [GHR92, Loc93].

Furthermore, we present an extension of higher-order patterns with decidable unification and another result that is tailored for the unification of induction schemes with first-order terms. It is shown that the unification of restricted second-order terms with first-order terms is decidable, where the restriction is such that typical induction schemes can be expressed. An example is the formula $\forall x.P(x) \Rightarrow P(x+1)$ in the inductive axiom

$$P(0), \forall x.P(x) \Rightarrow P(x+1) \vdash \forall x.P(x).$$

With these results only few classes remain where decidability of second-order unification is unknown.

2.8 Narrowing: The Higher-Order Case

The main contribution of this work is to lift several ideas of first-order narrowing to the higher-order case. Our main focus is lazy narrowing, which we develop first for full equational reasoning and then for functional-logic programming in Chapter 6. In addition, we consider other approaches to higher-order narrowing based on first-order ideas in Chapter 7.

In first-order narrowing, values for logic or free variables are computed via unification. The variables range over objects of the domains of interest. In a higher-order setting, unification can compute values for functional objects. For instance, a solution $F \mapsto \lambda x.x$ for the free variable F in the goal

$$\lambda x.diff(\lambda x.sin(F(x)),x) \to^? \lambda x.cos(x)$$

can be computed by narrowing. Apart from examples that intrinsically contain λ-terms, such as mathematics or programs, there are several other nice applications. For instance, we can handle quantifiers correctly and solve goals with mixed quantifier prefixes, such as

$$\forall x.\exists y.x =^? y.$$

Other examples can be found in Chapter 8.

We develop several optimizations and refinements for lazy narrowing. First, we focus on the general case, where we only assume that the rewrite rules are terminating. This permits the particularly important restriction to R-normalized solutions in order to limit narrowing steps. We also examine deterministic simplification on goals, which corresponds to (partial) functional evaluation. For instance, the above goal can be simplified by rewriting to obtain the new goal

$$\lambda x.cos(F(x)) * diff(\lambda x.F(x),x) \to^? \lambda x.cos(x).$$

Next, we narrow with the rule $X * 1 \to X$, which gives the new goals

$$\lambda x.cos(F(x)) \to^? X, diff(\lambda x.F(x),x) \to^? 1, X \to^? \lambda x.cos(x).$$

Now it is tempting to perform variable elimination, which means binding X to the term t for a goal $X \to^? t$ or $t \to^? X$. In our setting, deterministic elimination is complete for goals of the form $X \to^? t$. Thus it is safe to attempt only elimination of X in the third goal, which yields

$$\lambda x.cos(F(x)) \to^? \lambda x.cos(x), diff(\lambda x.F(x),x) \to^? 1.$$

Now the first goal can be solved by the higher-order unification rules, which compute the only solution $F \mapsto \lambda x.x$. After substituting this in the second goal, this remaining goal can be solved easily by evaluation for *diff*.

To complement the development of lazy narrowing, some alternative approaches to higher-order narrowing are discussed in Chapter 7. An overview of the different approaches to higher-order narrowing can be found in Figure 2.2. Our main approach, lazy narrowing, can be lifted to the higher-order case and is developed in Chapter 6 for both conditional and unconditional rewrite rules. For plain narrowing, which attempts to lift rewrite steps somewhere inside a term, we show that there are problems in the full higher-order

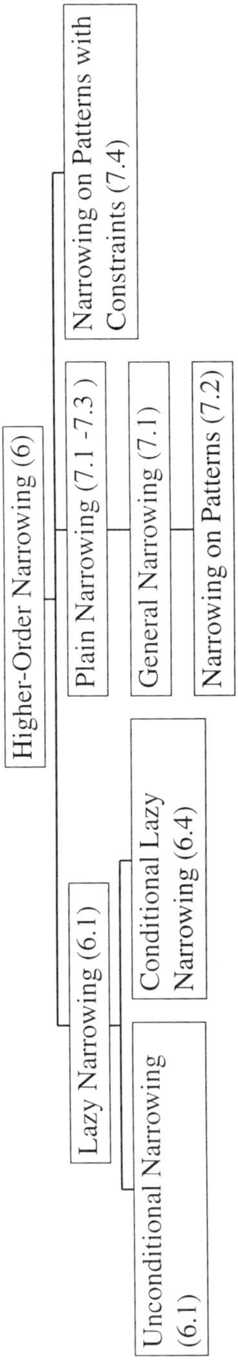

Figure 2.2: A Framework for Higher-Order Narrowing

case. We develop first a quite general notion, called General Narrowing in Section 7.1, which we use for examining the approach. For the restricted setting of higher-order patterns, this approach is however sufficient, as shown in Section 7.2. Finally, we explore in Section 7.4 an approach with combines the two other ones.

2.8.1 Functional-Logic Programming

Our approach to higher-order programming via narrowing is oriented towards functional languages. Recall that the core of modern functional languages such as SML [MTH90] and Haskell [PHA$^+$96] can be seen as higher-order rewrite rules. Our calculus for functional-logic programming is based on the results for lazy narrowing with terminating rules and is thus developed in Section 6.3.

Functional-logic programming can be viewed as a special case of narrowing with left-linear rewrite rules. Left-linearity means that free variables do not occur repeatedly on the left side of a rule. This small and common restriction has a significant impact on the expressiveness, i.e. full higher-order equality is avoided and decidable unification is gained. In addition, it is an important source of optimizations for narrowing and, most importantly, leads to the call-by-need strategy. This strategy generalizes the known call-by-need strategies of functional programming [Wad71, Lau93, AFM$^+$95].

The important observation is that in such a setting a particular class of goals, called Simple Systems, suffices for left-linear rules. The main invariants are that no variable occurs twice on the right-hand sides of a system of goals and that there are no cyclic dependencies between variable occurrences in goals. Simple Systems enjoy several nice properties. For instance, a variable cannot occur on both sides of a goal, e.g. $X \rightarrow^? f(X)$ is impossible and thus the occurs check is immaterial. Furthermore, solved forms are easy to detect. For instance, a Simple System of the form

$$t_1 \rightarrow^? X_1, \ldots, t_n \rightarrow^? X_n,$$

is guaranteed to have a solution. It follows from the invariants of Simple Systems that all $X_1, \ldots, X_n$ are distinct.

In Simple Systems a solution to the variable elimination problem leads to a new strategy, named call-by-need narrowing. Recall that variable elimination just binds a variable X in a goal $X \rightarrow^? t$ or $t \rightarrow^? X$ to the term t. For narrowing with terminating rules, variable elimination is deterministic for goals of the form $X \rightarrow^? t$. Elimination for the remaining case is undesirable in Simple Systems and, furthermore, such goals can safely be delayed. This simple idea leads to call-by-need narrowing, which computes values only if needed and also avoids copying. It generalizes call-by-need or lazy evaluation with sharing of identical subterms from the functional world to functional-logic programming.

The restriction to left-linear rewrite rules has one important consequence. Equality between arbitrary higher-order terms is not possible, as the usual encoding via a rule $X =^? X \to true$ is not permitted. Instead, one uses the so-called strict equality between first-order data types, which can still be expressed with left-linear rules. For programming applications, this is a sensible choice, since equality between functional expressions is expensive to compute and hence not desired.

Many higher-order extensions of functional-logic languages have been developed [BG86, She90, CKW89, GHR92, Loc93, Llo94, GHR97, NMI95, AKLN87]. To our knowledge, however, all of these are not complete w.r.t. higher-order equational logic. Some currently developed languages pursue similar goals, most notably Escher [Llo94] and Curry [HKMN95]. In both of these, functional evaluation is stressed over equational reasoning and completeness in a higher-order setting is not pursued. Other approaches treat higher-order variables as standing for exactly one function symbol [GHR97, NMI95], called applicative (higher-order)rewriting, and often employ a different underlying logic.

2.8.2 Conditional Narrowing and Higher-Order Programming

Another important ingredient of functional-logic programming are conditional rules, which often allow convenient specifications. For narrowing with conditional rules, we present a novel approach, which is only suitable in the higher-order case. We consider *normal conditional rules* of the form

$$l \to r \Leftarrow l_1 \to r_1, \ldots, l_n \to r_n,$$

where $l_i \to r_i$ denote conditions for the application of the rule and r_i are ground terms (i.e. without free variables) in R-normal form.

Although most first-order approaches are less restrictive, we argue that such extensions are not needed in a higher-order setting. Furthermore, this restriction has a significant advantage: for proving conditions of rules, as well as for queries, we consider *oriented* goals of the form $s \to^? t$ with solutions $\theta s \xrightarrow{*} t$. Hence this restriction permits a simpler operational model and is powerful enough to encode functional and logic programs: the core of modern functional languages can be seen as higher-order (unconditional) rewrite rules. Encoding logic programs is straightforward, as shown in Section 2.3, since the right-hand sides in the conditions are simply the constant *true*.

The restriction to ground right-hand sides is too strong for first-order functional-logic languages, as variables on the right in conditions serve as local variables. Consider for instance the function *unzip*, cutting a list of pairs into a pair of lists. In a functional language *unzip* can be written as

$$\begin{array}{lll} unzip([pair(X,Y)|R]) & \to & \mathsf{let}\ pair(xs,ys) \ = \ unzip(R) \\ & & \mathsf{in}\ pair([X|xs],[Y|ys]) \\ unzip([]) & \to & pair([],[]) \end{array}$$

where $pair(a,b)$ denotes a pair. The let-construct for pairs, written in mixfix notation as common in functional languages, can be defined a higher-order rewrite rule using β-reduction (see Section 8.1). In first-order functional-logic programming this function may be written as

$$\begin{array}{lll} unzip([pair(X,Y)|R]) & \to & pair([X|Xs],[Y|Ys]) \\ & & \quad \Leftarrow unzip(R) \to pair(Xs,Ys) \\ unzip([]) & \to & pair([],[]) \end{array}$$

The first of the above conditional rewrite rules has extra variables on the right, which are used to model the let-construct.

Notice that we permit new variables in the left sides of the conditions, which are used as "existential" variables, to be computed by unification as in logic programming. Consider for instance the following example modeling family relations, where a new variable Z is used in the definition of *grand_mother*. For brevity, we write just p for a rule $p \to true$ or a goal $p \to^? true$.

$$\begin{array}{l} mother(jane,mary) \\ mother(susan,mary) \\ mother(mary,judy) \\ wife(john,jane) \end{array}$$

$$grand\_mother(X,Y) \ \Leftarrow \ mother(X,Z), mother(Z,Y)$$

In the higher-order case, the concept of family relations can be generalized, similar to [Llo94, Nad87]:

$$\begin{array}{lll} family\_rel(wife) & & \\ family\_rel(mother) & & \\ family\_rel(join(R_1,R_2)) & \Leftarrow & family\_rel(R_1), family\_rel(R_2) \\ join(R_1,R_2,X,Y) & \Leftarrow & R_1(X,Z), R_2(Z,Y) \end{array}$$

In the last rules, *join* is intended to compose two relations. Thus a query

$$family\_rel(R), R(jane,judy)$$

should be answered by $R \mapsto join(mother, mother)$. Observe in this example that we can show properties of functional objects. Also partial application of *join* is useful in the above. For instance, the above rule could also be written as $join(R_1, R_2)(X, Y) \Leftarrow \ldots$.

We argue that many programming concepts are not only simpler expressed by higher-order functional programming, but also the technical treatment can be simpler. Handling extra variables for narrowing is both difficult and error-prone. For an overview of the extensive literature see [MH94]. Furthermore, there are several works [BG89, LS93, ALS94b] on confluence and termination of logic programs that correspond to such function constructs (sometimes called well-moded programs). For the termination of logic programs, such local variables are one of the main problems [dSD94]. Also, functional programming provides more directionality than logic programs, which is another major problem for proving termination [dSD94].

Chapter 3

Preliminaries

Basic definitions and results for higher-order equational reasoning are introduced in this chapter. The first sections contain general background material on reductions and orderings, followed by a brief introduction to λ-calculus. For a comprehensive treatment we refer to [HS86, Bar84].

3.1 Abstract Reductions and Termination Orderings

An **abstract reduction** is a relation on some set A. The following properties of reductions will be used mostly for term rewriting, which is a reduction on terms.

Definition 3.1.1 For some abstract reduction $\longrightarrow$, let $\longrightarrow^+$ denote its transitive closure, $\stackrel{*}{\longrightarrow}$ its reflexive transitive closure, and $\longleftarrow$ its inverse. Furthermore, define $\longleftrightarrow = \longrightarrow \cup \longleftarrow$. We write $\longrightarrow^n$ for some reduction of length n, i.e. $s_0 \longrightarrow^n s_n$ stands for a sequence $s_0 \longrightarrow s_1 \longrightarrow \ldots \longrightarrow s_n$.

A relation is an **equivalence relation** if it is reflexive, transitive and symmetric. A **partial ordering** is a reflexive, transitive and anti-symmetric relation. A **strict partial ordering** is a transitive and irreflexive relation.

A partial ordering $\leq$ is a **total ordering** if $a \leq b$ or $b \leq a$ holds for all a and b. A partial or total ordering $\leq$ is **compatible** with another partial ordering $\leq'$ if $\leq' \subseteq \leq$.

Definition 3.1.2 An abstract reduction is called **terminating** if no infinite reduction exists. An element a is called a **normal form** if no reduction from a exists.

Two elements s and t are **joinable** by a reduction $\longrightarrow$, written as $s{\downarrow}t$, if there exists u with $s \stackrel{*}{\longrightarrow} u$ and $t \stackrel{*}{\longrightarrow} u$. A reduction is called **locally confluent**, if any two reductions from an element t are joinable, i.e. if $t \longrightarrow u$ and $t \longrightarrow v$ then $u{\downarrow}v$. It is called **confluent**, if $\stackrel{*}{\longrightarrow}$ is locally confluent, i.e. if $t \stackrel{*}{\longrightarrow} u$ and $t \stackrel{*}{\longrightarrow} v$ then $u{\downarrow}v$.

Definition 3.1.3 The **lexicographic combination** of two reductions $\longrightarrow_1$ and $\longrightarrow_2$ on sets A and B, written as $R = (\longrightarrow_1, \longrightarrow_2)_{lex}$, is a reduction on $A \times B$, with $(a,b)\ R\ (a',b')$ if

- $a \longrightarrow_1 a'$ or
- $a = a'$ and $b \longrightarrow_2 b'$.

The important property of the lexicographic combination is the following:

Lemma 3.1.4 *The lexicographic combination of terminating reductions is terminating.*

The lexicographic combination of n abstract reductions $R^n = \longrightarrow_n, \ldots, \longrightarrow_1$ is defined recursively as $R^n_{lex} = (\longrightarrow_n, R^{n-1}_{lex})_{lex}$.

A **multiset** M over a set A is a mapping from A to $\{0,1,2,\ldots\}$. A multiset M can be viewed as a set where repeated elements are allowed, i.e. M maps an element $a \in A$ to its number of occurrences. A multiset M is finite if $M(x) > 0$ holds only for finitely many $x \in A$.

Removing an element from a multiset reduces the number of occurrences by one, if it occurs at all. Formally, removing an element a from a multiset M gives a new mapping $M' = M - a$ with $M'(x) = M(x)$ if $x \neq a$ and

$$M'(a) = \begin{cases} M(a) - 1 & \text{if } M(a) > 0 \\ 0 & \text{if } M(a) = 0 \end{cases}$$

Removing a multiset M from M', written as $M' - M$, is defined as the result of removing each occurrence of an element of M from M'. Adding an element to a multiset, written as $a + M$, and the union $M \cup M'$ of two multisets are defined correspondingly.

An important method for termination proofs is to extend an ordering $\ll$ on a set A to multisets of A. A **multiset** N **is smaller** than M, written as $N \ll_{multi} M$, if it can be obtained by removing one element from M plus adding finitely many smaller elements. Formally we have:

$$N \ll_{multi} M \quad \Leftrightarrow \quad \exists x \in A. M - x = N \cup N',$$

where N' is a finite multiset with $n \ll x, \quad \forall n \in N'$. As usual, the transitive closure of $\ll_{multi}$ will be used in most cases.

The following result allows one to extend termination orderings to multisets:

Theorem 3.1.5 ([DM79]) *The multiset extension of a terminating ordering is terminating.*

Besides (multi-)sets, we often use **lists,** which are denoted by square brackets, i.e. appending a list R to an element t is written as $[t|R]$. The application of a function f to a list, written as $f[\overline{t_n}]$, is defined as $[\overline{f(t_n)}]$.

3.2 Higher-Order Types and Terms

This section introduces our term language: simply typed λ-terms. The set of types $\mathcal{T}$ for the simply typed λ-terms is generated by a set $\mathcal{T}_0$ of **base types** (e.g. int, bool) and the **function type constructor** $\rightarrow$. Notice that $\rightarrow$ is right associative, i.e. $\alpha \rightarrow \beta \rightarrow \gamma = \alpha \rightarrow (\beta \rightarrow \gamma)$. We assume a set of **variables** V_τ, and a set of **constants** C_τ for all types $\tau \in \mathcal{T}$, where $V_\tau \cap V_{\tau'} = C_\tau \cap C_{\tau'} = \{\}$. The set of all variables is $V = \bigcup_{\tau \in \mathcal{T}} V_\tau$, which is disjoint from the set of all constants, $C = \bigcup_{\tau \in \mathcal{T}} C_\tau$. **Atoms** are either constants or variables. The following **naming conventions** are used in the sequel:

- F, G, H, P, X, Y free variables,
- a, b, c, f, g (function) constants,
- x, y, z bound variables,
- α, β, τ type variables.

Further, we often use s and t for terms and u, v, w for constants or bound variables. The following grammar defines the syntax for **untyped λ-terms**

$$t \;=\; F \mid x \mid c \mid \lambda x.t \mid (t_1\ t_2),$$

where $(t_1\ t_2)$ denotes the **application** of two terms. The term $\lambda x.t$ denotes an **abstraction** over x and thus creates a new functional object. An occurrence of a variable x in a term t is **bound**, if it occurs below a binder for x, i.e. the occurrence of x is in a subterm $\lambda x.t'$. Otherwise it is a **free** occurrence. . Free and bound variables of a term t will be denoted as $\mathcal{FV}(t)$ and $\mathcal{BV}(t)$, respectively.

Notice that there can be many such binders, e.g. $\lambda x.\lambda x.x$, but only the innermost one is associated with x. To avoid such cases, we will adopt assumptions (see below) on the naming of bound variables for simplicity.

A list of syntactic objects $s_1, \ldots, s_n$ where $n \geq 0$ is abbreviated by $\overline{s_n}$. We will use n-fold abstraction and application, written as $\lambda \overline{x_n}.s = \lambda x_1 \ldots \lambda x_n.s$ and $a(\overline{s_n}) = ((\cdots(a\ s_1)\cdots)\ s_n)$, respectively. For instance

$$\lambda \overline{x_m}.f(\overline{s_n}) = \lambda x_1 \ldots \lambda x_m.((\cdots(f\ s_1)\cdots)\ s_n)$$

A **type judgment** stating that t is of type τ is written as $t : \tau$. The following inference rules inductively define the set of **simply typed λ-terms**.

$$\frac{x \in V_\tau}{x : \tau} \qquad \frac{c \in C_\tau}{c : \tau}$$

$$\frac{s : \tau \to \tau' \quad t : \tau}{(s\ t) : \tau'} \qquad \frac{x : \tau \quad s : \tau'}{(\lambda x.s) : \tau \to \tau'}$$

The **order of a type** $\varphi = \alpha_1 \to \ldots \to \alpha_n \to \beta,\ \ \beta \in \mathcal{T}_0$ is defined as

$$Ord(\varphi) = \begin{cases} 1 & \text{if } n = 0, \text{ i.e. } \varphi = \beta \in \mathcal{T}_0 \\ 1 + k & \text{otherwise, where} \\ & k = max(Ord(\alpha_1), \ldots, Ord(\alpha_n)) \end{cases}$$

We say a symbol is of order n if it has a type of order n. A **term of order n** is restricted to

- function constants of order $\leq n+1$ and
- variables of order $\leq n$.

For instance, if a term $F(\overline{t_n})$ is second-order, then all subterms t_i must be of base type. We say a term t is **weakly second-order** if it is second-order, but with the exception that bound variables of arbitrary type may occur as arguments to free variables (up to η-equality). For instance, $F(\lambda z.x(z), y)$ is weakly second-order, but not second-order.

Let $\{x \mapsto s\}t$ denote the result of replacing every free occurrence of x in t by s. Note that we assume that free variables in s are not captured by substitution, which means they are not substituted into the scope of a bound variable of the same name. To avoid this, we assume that bound variables are renamed appropriately, as in the following definition of α-conversion. (In the further treatment, our variable conventions will avoid such cases.) A detailed definition can for instance be found in [HS86].

The **conversions in λ-calculus** are defined as:

- α-conversion: $\lambda x.t \succ_\alpha \lambda y.(\{x \mapsto y\}t)$, if $y \notin \mathcal{FV}(t)$
- β-conversion: $(\lambda x.s)t \succ_\beta \{x \mapsto t\}s$
- η-conversion: $\lambda x.(t\,x) \succ_\eta t$, if $x \notin \mathcal{FV}(t)$

The first of the above, α-conversion, serves for renaming bound variables. The second rule, β-conversion, replaces the formal parameter of a function $\lambda x.s$ by the argument t. The rule of η-conversion entails extensionality, which means that two functions are considered equal, if they behave equally for all arguments. Formally, **extensionality** is defined as $\forall x.f(x) = g(x) \Rightarrow f = g$.

A **β-redex** is a term of the form $(\lambda x.s)t$ where β-reduction applies, and similarly for the other reductions. For $\phi \in \{\alpha, \beta, \eta\}$ we write $s \rightarrow_\phi t$, called **ϕ-reduction**, if t is obtained from s by ϕ-conversion on some subterm of s. Let $\rightarrow_{\beta,\eta}$ be defined as $\rightarrow_\beta \cup \rightarrow_\eta$, and similarly for other combinations. The reflexive, symmetric and transitive closure of some Φ-reduction induces an equivalence relation on terms, written as $s =_\Phi t$, where $\Phi \subseteq \{\alpha, \beta, \eta\}$. Application of the conversion rules in the other direction is called **expansion**. Reduction in the simply typed λ-calculus is confluent and terminating w.r.t. β-reduction (and w.r.t. η-reduction), see e.g. [Bar84]. Hence, to check $s =_{\beta,\eta} t$, it suffices to compare the normal forms of the two terms.

The **β-normal form** (**η-normal form**) of a term t is denoted by $t{\downarrow_\beta}$ ($t{\downarrow_\eta}$). Let t be in β-normal form. Then t is of the form $\lambda\overline{x_n}.v(\overline{u_m})$, where v is called the **head** of t, and written as $Head(t)$. The **η-expanded form** of a term $t = \lambda\overline{x_n}.v(\overline{u_m})$ is defined by

$$t{\uparrow_\eta} = \lambda\overline{x_{n+k}}.v(\overline{u_m{\uparrow_\eta}}, x_{n+1}{\uparrow_\eta}, \ldots, x_{n+k}{\uparrow_\eta})$$

where $t : \overline{\tau_{n+k}} \rightarrow \tau$ and $x_{n+1}, \ldots, x_{n+k} \notin \mathcal{FV}(\overline{u_m})$. We call $t{\downarrow_\beta}{\uparrow_\eta}$ the **long $\beta\eta$-normal form** of a term t, also written as $t{\updownarrow_\beta^\eta}$. A term t is in long $\beta\eta$-normal form if $t = t{\updownarrow_\beta^\eta}$. It is well known [HS86] that $s =_{\alpha\beta\eta} t$ iff $s{\updownarrow_\beta^\eta} =_\alpha t{\updownarrow_\beta^\eta}$.

The **size** $|t|$ of a term t in long $\beta\eta$-normal form is defined as the number of symbols occurring in t, not counting binders λx:

$$\begin{array}{lcl} |\lambda x.t| & = & |t|, \\ |s\,t| & = & |s| + |t|, \\ |v| & = & 1, \quad v \in V \cup C \end{array}$$

A variable is **isolated** if it occurs only once (in a term or in a system of equations). A term is **linear** if no free variable occurs repeatedly. A term $\lambda\overline{x_k}.v(\overline{t_n})$ is called **flexible** if v is a free variable and **rigid** otherwise.

Assumptions and Conventions

We will in general assume that terms are in long $\beta\eta$-normal form. For brevity, we write variables in η-normal form, e.g. X instead of $\lambda\overline{x_n}.X(\overline{x_n})$. We assume that the transformation into long $\beta\eta$-normal form is an implicit operation.

We work in the following completely modulo α-conversion, which means that α-equivalent terms are identified. There exist representations for λ-terms which achieve α-conversion on a syntactical basis. For instance, with de Bruijn indices [dB72], bound variables are represented as natural numbers, indicating the corresponding binder. The main result is that two α-equivalent terms have the same de Bruijn representation.

We follow the **variable convention** that free and bound variables are kept disjoint (see also [Bar84]). We cannot enforce this convention completely. For instance, in the congruence rule used in Section 4.3.1, $s = t \Rightarrow \lambda x.s = \lambda x.t$, x occurs both free and bound. More seriously, this distinction permits so-called **loose bound** variables, i.e. "bound" variables without a binder. Such variables are typically created when a subterm of a term is considered or manipulated. For instance, $f(x)$ is a subterm of $\lambda x.g(f(x))$ with a loose bound variable. In such cases, these variables can be viewed as bound variables where the binder is (implicit) in the context. In general, loose bound variables may create inconsistencies. Although sometimes convenient, we will avoid loose bound variables whenever possible.

For simplicity, we assume that bound variables with different binders have different names. As a consequence of our conventions, it suffices to write $s = t$ instead of $s =_{\alpha,\beta,\eta} t$, as we assume long $\beta\eta$-normal form and work modulo α-conversion. These conventions for instance permit the following definition. We say a bound variable y in a term $\lambda\overline{x_n}.t$ in long $\beta\eta$-normal form is **outside bound** if $y = x_i$ for some i.

3.3 Positions in λ-Terms

We describe positions in λ-terms by sequences over natural numbers, as we have adopted n-ary application. Such a sequence describes the **path** to a subterm of a term. Positions in λ-terms are often written as sequences over 1 and 2. It is easy to translate one representation into the other, as in the following example for the term $\lambda x_1, x_2.F(a,b)$ in Figures 3.1 and 3.2. Notice that our representation of terms as trees is a generalization of usual first-order terms and positions.

Let ε denote the **empty sequence**, let $i.p$ denote the sequence p appended to an element i, and let $p + p'$ concatenate two sequences. A sequence p is a **prefix** of p', if $\exists q.p + q = p'$, and similarly p is a **postfix** if $\exists q.q + p = p'$.

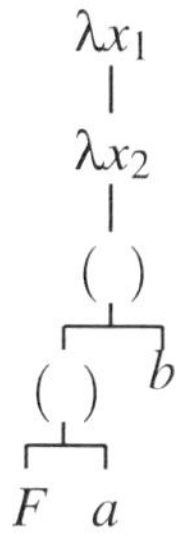

Figure 3.1: $\lambda x_1, x_2.F(a,b)$ as Binary Tree

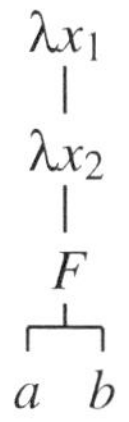

Figure 3.2: $\lambda x_1, x_2.F(a,b)$ as n-nary Tree

Definition 3.3.1 The **subterm** of s at **position** p, written as $s|_p$, is defined as

- $s|_\varepsilon = s$
- $v(\overline{t_m})|_{i.p} = t_i|_p$ if $i \leq m$
- $\lambda\overline{x_m}.t|_{1.p} = (\lambda x_2, \ldots, x_m.t)|_p$
- undefined otherwise

The following notion of subterm extends the definition of subterms to account for a binding environment. A term $s = \lambda\overline{x_n}.s_0$ is a **subterm modulo binders** of $t = \lambda\overline{x_n}.t_0$, written as $s <_{sub} t$, if s_0 is a (true) subterm of t_0.

A term t with the subterm at position p replaced by s is written as $t[s]_p$. Two positions p and q are **independent** if none is a prefix of the other. For a term s of the form $\lambda\overline{x_k}.v(\overline{t_n})$, the position of v is called the **root position**. A (sub-)term $t|_p$ is called **ground** if no free variables of t occur in $t|_p$. Note that this is well defined as we syntactically distinguish between bound and free variables.

If p is a position in s then let $\mathcal{BV}(s,p)$ be the set of all λ-abstracted variables on the path to p in s. Such a path is called **rigid** if it contains no free variables.

3.4 Substitutions

Substitutions are finite mappings from variables to terms, which are denoted by $\{\overline{X_n \mapsto t_n}\}$ and extend homomorphically from variables to terms. In general, substitutions map only the free variables of a term. If $s = \theta t$ for some substitution θ, then s is called an **instance** of t.

We define the domain of a substitution as $\mathcal{D}om(\theta) = \{X \mid \theta X \neq X\}$, the image as $Im(\theta) = \{\theta X \mid X \in \mathcal{D}om(\theta)\}$, and the range as $\mathcal{R}ng(\theta) = \mathcal{FV}(Im(\theta))$. The **free variables of a substitution** θ are defined as $\mathcal{FV}(\theta) = \mathcal{D}om(\theta) \cup \mathcal{R}ng(\theta)$. For a list of syntactical objects $\overline{C_n}$ we write $\mathcal{FV}(\overline{C_n})$ instead of $\mathcal{FV}(C_1) \cup \ldots \cup \mathcal{FV}(C_n)$. Two **substitutions are equal** on a set of variables W, written as $\theta =_W \theta'$, if $\theta X = \theta' X$ for all $X \in W$. The **restriction of a substitution** to a set of variables W is defined by $\theta|_W X = \theta X$ if $X \in W$ and $\theta|_W X = X$ otherwise. The **composition** $\delta\theta$ of two substitutions is defined as $(\delta\theta)(s) = \delta(\theta(s))$.

Definition 3.4.1 A substitution θ is **more general** than θ' over a set of variables W, written as $\theta' \leq_W \theta$, if $\theta' =_W \sigma\theta$ for some substitution σ.

For brevity, we will often leave the set of variables W implicit and write $\theta' \leq \theta$ or $\theta = \theta'$. A substitution θ is **idempotent** iff $\theta = \theta\theta$. We will in general assume that substitutions are idempotent. This is justified by the following two basic lemmata [SG89].

Lemma 3.4.2 *A substitution θ is idempotent if $\mathcal{R}ng(\theta) \cup \mathcal{D}om(\theta) = \{\}$.*

In the higher-order order-case, this condition for idempotence is only sufficient but not necessary, as noted in [SG89].

Lemma 3.4.3 *For any substitution θ and set of variables W with $\mathcal{D}om(\theta) \subseteq W$, there exists an idempotent substitution θ' such that $\mathcal{D}om(\theta) = \mathcal{D}om(\theta')$, $\theta' \leq \theta$ and $\theta \leq_W \theta'$.*

As we syntactically distinguish between bound and free variables, we can speak of **well-formed** substitutions: a substitution is well-formed, if it does not contain loose bound variables, i.e. bound variables without binder. With a few exceptions, we will in general assume well-formed substitutions. Thus, for instance, $\theta\lambda\overline{x_k}.t = \lambda\overline{x_k}.\theta t$ by convention.

Properties of terms extend to substitutions in the component-wise way, i.e. to the terms in the image. For instance, a substitution θ is ground (in long $\beta\eta$-normal form) if all terms in the image of θ are ground (in long $\beta\eta$-normal form). Let $\rightarrow$ be an arbitrary relation on terms. We define $\theta \rightarrow \theta'$ to mean that $\theta F \rightarrow \theta' F$ holds for all $F \in \mathcal{Dom}(\theta)$. Similarly, $\theta\!\downarrow$ denotes the substitution with $F \mapsto (\theta F)\!\downarrow$ for all $F \in \mathcal{Dom}(\theta)$.

3.5 Unification Theory

Unification of two terms s and t aims at finding a substitution θ such that $\theta s = \theta t$, where θ is called a **unifier** of s and t. Unification problems are written as $s =^? t$. For examples, we refer to the next section. There exist several surveys on the subject [BS94, JK91].

An **equational theory** E is an equivalence relation on terms that is stable under substitutions, i.e. $s =_E t$ implies $\theta s =_E \theta t$ for any substitution θ. Usually, equational theories are generated by a set of equations, as discussed for the higher-order case in Section 4.3.1. The conversions of λ-calculus, i.e. the $\alpha\beta\eta$-rules, are equations with meta-level conditions.

A substitution τ is **more general** than σ, modulo a theory E over a set of variables W, written as $\sigma \leq_{E,W} \tau$, if $\exists \delta . \sigma =_{E,W} \delta\tau$. Accordingly, $\sigma =_{E,W} \tau$ if $\sigma X =_E \tau X$ for all $X \in W$. For simplicity, we often leave the parameter W implicit.

Unification modulo an equational theory $=_E$, or E**-unification,** aims at finding a substitution with $\theta s =_E \theta t$. Then θ is called an E-unifier of s and t. As there can be many solutions to a unification problem $s =^? t$, it is desirable to find minimal sets of solutions, as defined next:

Definition 3.5.1 A set of substitutions S is a **minimal, complete set of unifiers (MCSU)** of a unification problem $s =^? t$ for some equational theory E, iff

- Each element of S is an E-unifier of $s =^? t$.
- For every E-unifier of $s =^? t$ there exists a more general E-unifier in S.
- The elements of S are incomparable.

It can be shown that such a set of incomparable common instances is uniquely defined except for variants of its elements [FH86]. There exist several classes of unification problems, depending on the existence of a MCSU. We call an E-unification problem

unitary if a MCSU is either empty or a singleton,

finitary if a finite MCSU exists,

infinitary if a possibly infinite MCSU exists,

nullary if no MCSU may exist.

This classification extends to an equational theory E, if for all E-unification problems the property (e.g. unitary) holds.

Another distinction of unification problems is sometimes considered in the first-order case [BS94]: are only constant symbols of a fixed signature allowed in the terms to be unified or arbitrary constants? In the first-order case, the above classification may depend on this. This distinction is immaterial in a higher-order context, as we must deal with local "constants", i.e. bound variables. For instance, assume an equational theory E for a symbol f. Then solving an equation $f(a) =_E f(X)$ with an extra constant a is equivalent to solving $\lambda x.f(x) =_E \lambda x.f(X(x))$. (It is well known that it is sufficient in λ-calculus to consider only bound variables, see for instance [Mil92].)

3.6 Higher-Order Patterns

The following subclass of λ-terms was introduced originally by Dale Miller [Mil91a] and is often called higher-order patterns in the literature.

Definition 3.6.1 A simply typed λ-term s in β-normal form is a **relaxed higher-order pattern**, if all free variables in s only have bound variables as arguments, i.e. if $X(\overline{t_n})$ is a subterm of s, then all $t_i{\downarrow}_\eta$ are bound variables.

Some examples of relaxed higher-order patterns are the terms $\lambda x, y.F(x,x,y)$ and $\lambda x.f(G(\lambda z.x(z)))$, where the latter is at least third-order. Non-patterns are $\lambda x, y.F(a,y)$, $\lambda x.G(H(x))$.

In most of the existing literature [Mil91b, Nip91a], patterns are required to have distinct bound variables as arguments to a free variable. This restriction is necessary for unitary unification, but for some of the results on decidability of higher-order unification in Chapter 5 this is not relevant.

Definition 3.6.2 A (**higher-order**) **pattern** is a relaxed pattern where the arguments to free variables are distinct bound variables.

For instance, $\lambda x, y.F(x,x,y)$ is a relaxed but not a higher-order pattern. An example for a higher-order pattern is $\lambda x, y.G(x, \lambda z.y(z))$. Unification of patterns is decidable and a most general unifier exists if they are unifiable [Mil91a,

Nip91a], as shown in Section 4.2. Furthermore, a most general unifier can be computed in linear time [Qia93]. This shows that unification with patterns behaves similar to the first-order case.

Several important properties of patterns with respect to term rewriting are examined later and are based on the important fact that β-reduction on patterns only renames bound variables. For this reason β-reduction on patterns is also called β_0-reduction in [Mil91a].

Chapter 4

Higher-Order Equational Reasoning

This chapter introduces higher-order unification and term rewriting. First, Section 4.1 reviews a set of transformation rules for full higher-order pre-unification. This is followed by an important special case, higher-order patterns, where unification proceeds almost as in the first-order case.

4.1 Higher-Order Unification by Transformation

We present in the following a version of the transformation system PT for higher-order unification of Snyder and Gallier [SG89]. More precisely, we adapt the primed transformations for pre-unification of Section 5 in [SG89].

Consider solving an equation $\lambda\overline{x_k}.F(\overline{t_n}) =^? \lambda\overline{x_k}.v(\overline{t'_m})$ where v is not a free variable. Such equations are called **flex-rigid**. Clearly, for any solution θ to F the term $\theta F(\overline{t_n})$ must have (after β-reduction) the symbol v as its head. There are two possibilities, where the first only occurs in the higher-order case.

- In the first case, v already occurs in (the solution to) some t_i. For instance, reconsider the equation $F(a) =^? a$, where $\{F \mapsto \lambda x.x\}$ is a solution based on a **projection**. In general, a projection binding for F is of the form $\{F \mapsto \lambda\overline{x_n}.x_i(\ldots)\}$. As some argument, here a, is carried to the head of the term, such a binding is called projection. This name was introduced in [JP76].

- The second case is that the head of the solution to F is just the desired symbol v. For instance, in the last example, an alternative solution is $\{F \mapsto \lambda x.a\}$. This is called **imitation.** Notice that imitation is not possible if v is a bound variable.

To solve a flex-rigid pair, the strategy is to guess an appropriate imitation or projection binding only for one rigid symbol, here a, and thus approximate the solution to F. Unification proceeds by iterating this process which focuses only on the outermost symbol. Roughly speaking, the rest of the solution for F is left open by introducing new variables, as shown formally in the next definition of these bindings.

Deletion

$$t =^? t \quad \Rightarrow \quad \{\}$$

Decomposition

$$\lambda\overline{x_k}.v(\overline{t_n}) =^? \lambda\overline{x_k}.v(\overline{t'_n}) \quad \Rightarrow \quad \{\overline{\lambda\overline{x_k}.t_n =^? \lambda\overline{x_k}.t'_n}\}$$

Elimination

$$F =^? t \quad \Rightarrow^{\theta} \quad \{\} \text{ if } F \notin \mathcal{FV}(t) \text{ and where } \theta = \{F \mapsto t\}$$

Imitation

$$\lambda\overline{x_k}.F(\overline{t_n}) =^? \lambda\overline{x_k}.f(\overline{t'_m}) \quad \Rightarrow^{\theta} \quad \{\overline{\lambda\overline{x_k}.H_m(\overline{\theta t_n}) =^? \lambda\overline{x_k}.\theta t'_m}\}, \text{ where } \theta = \{F \mapsto \lambda\overline{x_n}.f(\overline{H_m(\overline{x_n})})\} \text{ is an appropriate imitation binding}$$

Projection

$$\lambda\overline{x_k}.F(\overline{t_n}) =^? \lambda\overline{x_k}.v(\overline{t'_m}) \quad \Rightarrow^{\theta} \quad \{\lambda\overline{x_k}.\theta t_i(\overline{H_j(\overline{t_n})}) =^? \lambda\overline{x_k}.v(\overline{\theta t'_m})\}, \text{ where } \theta = \{F \mapsto \lambda\overline{x_n}.x_i(\overline{H_j(\overline{x_n})})\} \text{ is an appropriate projection binding}$$

Figure 4.1: System PT for Higher-Order Pre-Unification

Definition 4.1.1 Assume an equation $\lambda\overline{x_k}.F(\overline{t_n}) =^? \lambda\overline{x_k}.v(\overline{t'_m})$, where all terms are in long $\beta\eta$-normal form. An **imitation binding** for F is of the form

$$F \mapsto \lambda\overline{x_n}.f(\overline{H_m(\overline{x_n})})$$

where $\overline{H_m}$ are new variables of appropriate type. A **projection binding** for F is of the form

$$F \mapsto \lambda\overline{x_n}.x_i(\overline{H_p(\overline{x_n})})$$

where $\overline{H_p}$ are new variables with $\overline{H_p : \overline{\alpha_n} \to \tau_p}$ and $x_i : \overline{\tau_p} \to \tau$. A **partial binding** is an imitation or a projection binding.

Notice that in the above definition, the bindings are not written in long $\beta\eta$-normal form. The long $\beta\eta$-normal form of an imitation or projection binding can be written as

$$F \mapsto \lambda\overline{x_n}.v(\overline{\lambda\overline{z_{j_p}}.H_p(\overline{x_n},\overline{z_{j_p}})}).$$

A full exhibition of the the types involved can be found in [SG89].

The transformation rules PT for higher-order unification in Figure 4.1 consist of the basic rules for unification, such as Deletion, Elimination and Decomposition plus the two rules explained above: Imitation and Projection. The rules work on pairs of terms to be unified, written as $\{u =^? v, \ldots\}$, and return a set of new pairs. We extend the transformation rules on goals to sets of goals in the canonical way: $\{s =^? t\} \cup S \Rightarrow^\theta \overline{\{s_n =^? t_n\}} \cup \theta S$ if $s =^? t \Rightarrow^\theta \overline{\{s_n =^? t_n\}}$. We abbreviate a sequence of transformations

$$G_0 \Rightarrow^{\delta_1}_{PT} G_1 \Rightarrow^{\delta_2}_{PT} \cdots \Rightarrow^{\delta_{n-1}}_{PT} G_{n-1} \Rightarrow^{\delta_n}_{PT} G_n$$

by $\overset{*}{\Rightarrow}{}^{\delta}_{PT}$, where $\delta = \delta_n \ldots \delta_1$.

Notice that the rules in Figure 4.1 only perform pre-unification. Pre-unification differs from unification by the handling of so-called **flex-flex pairs.** These are equations of the form $\lambda\overline{x_k}.P(\ldots) =^? \lambda\overline{x_k}.P'(\ldots)$. Huet showed that such pairs of order three may not have a MCSU [Hue76]: there may exist an infinite chain of unifiers, one more general than the other, without any most general one. The important idea to remedy this situation is that flex-flex pairs are guaranteed to have at least one unifier, e.g. $\{P \mapsto \lambda\overline{x_m}.a, P' \mapsto \lambda\overline{x_n}.a\}$. The idea of pre-unification is to handle flex-flex pairs as constraints and not to attempt to solve them explicitly.

A substitution θ is a **pre-unifier** of s and t if the equation $\theta s =^? \theta t$ can be simplified by Deletion, Decomposition and Elimination to a set of flex-flex pairs. In other words, $\theta s =^? \theta t$ only differ at subterms that have variable heads. The notions of MCSU and unification classes in Section 3.5 extend straightforwardly to pre-unification.

In this work, we will often use the restriction to second-order terms. The only place where the restriction to second-order terms simplifies the system is the last rule, projection, where x_i must be of base type. Hence the binding to F in this case is of the simpler form $F = \lambda\overline{x_n}.x_i$, which will be important for our results. As we will often encounter this case, we give an explicit simplified rule:

Second-Order Projection

$$\lambda\overline{x_k}.F(\overline{t_n}) =^? \lambda\overline{x_k}.v(\overline{t'_m}) \quad \Rightarrow^\theta \quad \{\lambda\overline{x_k}.\theta t_i =^? \lambda\overline{x_k}.v(\overline{\theta t'_m})\}$$
$$\text{where } \theta = \{F \mapsto \lambda\overline{x_n}.x_i\}$$

The following soundness lemma is easy to show:

Theorem 4.1.2 *System PT is a sound transformation system for higher-order pre-unification.*

When applying the rules of system PT to a set of equations, there are two sources of non-determinism:

1. Which rule to apply
2. to which equation.

If it is possible for one rule application to disregard other rule applications without losing completeness, this application is called deterministic. It was shown in the early work by Huet [Hue76] that completeness does not depend on how the equations are selected. This is implicit in the proof in [SG89], and is also explained at the end of this section. Furthermore, the only branching occurs when both Imitation and Projection apply to some equation. In other words, application of the first three rules is deterministic [SG89].

There are several further optimizations. An example is stripping off a binder λx if x does not occur. For instance, assume an equation $\lambda x, y.P(x) =^? \lambda x, y.f(a)$, for which the Elimination rule does not apply. Clearly, the binder λy is superfluous here and can be removed. Then the elimination rule applies directly.

Example 4.1.3 Consider the unification problem at the root of the search tree in Figure 4.2, which is obtained by the transformations PT in Figure 4.1. Notice that in this example all projection bindings are of the form $\lambda x.x$. The failure cases are caused by a clash of distinct symbols and are abbreviated. The partial bindings of the successful path yield the only solution $\{F \mapsto \lambda x.g(x,x), G_1 \mapsto G, G_2 \mapsto G\}$.

The following general result on higher-order unification will be important for results in the further sections and can e.g. be found in [Sny91].

Lemma 4.1.4 *If θ is a maximally general pre-unifier of an equation $s =^? t$, then $\mathcal{Dom}(\theta) \subseteq \mathcal{FV}(s,t)$.*

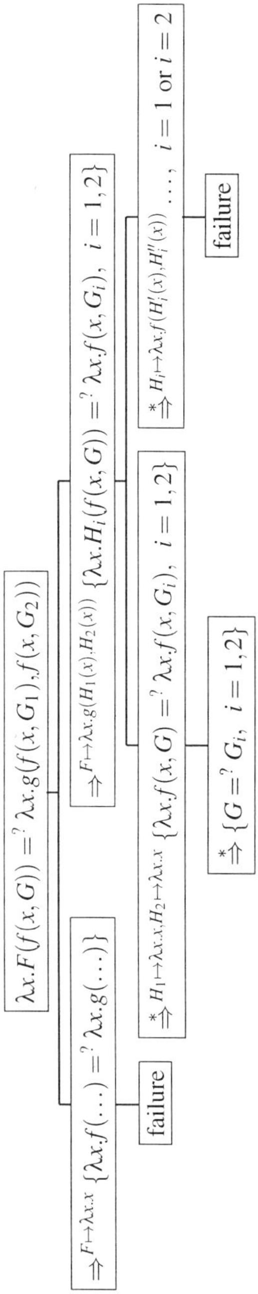

Figure 4.2: Search Tree with System PT

It should be mentioned that the Elimination rule is not needed for completeness: any equation of the form $P =^? \lambda\overline{x_k}.t$ where Elimination applies can be solved by repeated Imitation and Projection, until only flex-flex pairs remain. The only difference is that more flex-flex pairs may remain, as the Elimination rule also applies to some of such pairs. We sometimes use the restriction that Elimination is not applied to flex-flex pairs, which is sufficient for decidability results. The same is done in the algorithm presented by Snyder et al. [SG89].

This leads to another interesting observation: in contrast to Elimination, Imitation and Projection only compute substitutions that map terms to higher-order patterns. Composing pattern substitutions again yields pattern substitutions.

Fact 4.1.5 *For higher-order pre-unification it is sufficient to consider pattern substitutions.*

Intuitively, this holds since for pre-unification terms only have to agree at non-variable positions.

Completeness of Higher-Order Unification

We sketch in the following the completeness result for PT along the lines of [SG89], where the full treatment can be found. We say that a substitution δ **approximates** a substitution θ for a variable F if there exists a substitution θ' with

- $\mathcal{Dom}(\theta') = \mathcal{Dom}(\theta) - \{F\} \cup \mathcal{Rng}(\delta)$
- $\theta F = \theta' \delta F$
- $\theta =_W \theta'$, where $W = \mathcal{Dom}(\theta) - \{F\}$.

The following results are adapted from [SG89]. The next lemma shows that partial bindings approximate solutions.

Lemma 4.1.6 *For any flex-rigid equation $\lambda\overline{x_k}.F(\overline{t_n}) =^? \lambda\overline{x_k}.v(\overline{t'_m})$ with solution θ there exists a partial binding δ for F such that δ approximates θ.*

Theorem 4.1.7 (Completeness of PT) *If $s =^? t$ has solution θ, i.e. $\theta s = \theta t$, then $\{s =^? t\} \overset{*}{\Rightarrow}{}^{\delta}_{PT} F$ such that δ is more general than θ and F is a set of flex-flex goals.*

Proof The proof proceeds by induction on the following lexicographic termination ordering on $(\overline{E_n}, \theta)$, where for $\overline{E_n}$ is a system of equations with solution θ.

- A: compare the multiset of sizes of the bindings in θ, if equal
- B: compare the multiset of sizes of the equations $\overline{E_n}$.

Notice that a transformation not only changes $\overline{E_n}$, but also the associated solution has to be updated. That is, in case of a binding $F \mapsto t$, the variable F is removed from θ, and, if it is a partial binding, solutions for the new variables in t are added.

If E is in solved form, nothing remains to show. Otherwise, select some non flex-flex equation from $\overline{E_n}$. It is trivial to see that at least one transformation must apply. For each case we show that the ordering is reduced and that the solution is approximated. If the equation is a trivial pair $s =^? s$, ordering B is reduced. In case of the Decomposition rule B is reduced.

When eliminating an equation $F =^? t$, the binding $\{F \mapsto t\}$ clearly approximates θ as $\theta F = \theta t$. Since the new solution contains fewer bindings, A is reduced.

For the Imitation and Projection rule consider an equation $\lambda \overline{x_k}.F(\overline{t_n}) =^? t$. In this case, Lemma 4.1.6 shows that there exists a partial binding δ that approximates θ with θ', i.e. $\theta = \theta'\delta$. Furthermore, all new bindings in θ' for the new variables in δF are smaller than the binding for F in θ, thus reducing measure A. □

It is easy to see from the recursive structure of the last proof that the completeness does not depend on the selection of goals. Each subgoal is solvable independently, or it is a flex-flex equation. In contrast to the first-order case, the selection is more limited, as flex-flex goals are delayed. Notice that flex-flex goals can become non flex-flex pairs by instantiation.

4.2 Unification of Higher-Order Patterns

Unification of higher-order patterns is a special case of higher-order unification that proceeds similar to first-order unification. The main advantage is that most general unifiers exist for patterns. Compared to higher-order unification, there is no choice between Projection and Imitation. Only the flex-flex cases are more involved than the first-order case. Using efficient data structures, Qian [Qia93] showed that a linear-time implementation of pattern unification is possible.

The following set of rules for unification of higher-order patterns is taken from [Nip93a]. The exposition there includes a rule that strips off binders, i.e.

$$\lambda x.s =^? \lambda x.t \Rightarrow s =^? t$$

This assumes that bound variables are distinguished syntactically and is in fact closer to an implementation, as working with full binders is rather tedious. Notice that the η-extended form is often not practical, in particular for variables, free or bound, of higher type. Then η-expansion has to be performed during unification.

The transformations in Figure 4.3 work on lists, as the order of application is important for termination [Nip93a]. The problem is that the algorithm introduces new variables on the way and repeating this eagerly may lead to non-termination. For instance, consider $[c(X) = Y, Y =^? X] \Rightarrow_{PU} [X =^? Y_1, c(Y_1) =^? X]$. Here the occurs check applies only after an elimination whereas repeated imitation diverges.

A different method for solving equations of the form $\lambda\overline{x_k}.P(\overline{y_n}) =^? t$ is presented in a more general context in Section 5.1. This method does not introduce temporary variables and is in fact closer to an implementation (e.g. [Nip93a]).

The algorithm coincides with standard first-order unification algorithms, e.g. [JK91], for first-order terms. Notice that in the Flex-Flex rules any permutation of the bound variables $\overline{z_p}$ is sufficient for computing a most general unifier.

Theorem 4.2.1 ([Mil91a, Nip93a]) *System PU computes a most general unifier for two higher-order patterns if a unifier exists.*

Although this algorithm introduces new variables, in contrast to its first-order companion, it has the following important property:

Lemma 4.2.2 *If θ is a most general unifier of two pattern s and t, then* $|\mathcal{FV}(\theta s)| \leq |\mathcal{FV}(s,t)|$.

Proof by induction on the length of PU reductions. □

This lemma and the following property give some insight on the variables introduced by PU and will be important for some termination proofs in Chapter 5.

A substitution θ is **size increasing**, if $|X(\overline{y_n})| < |\theta X(\overline{y_n})|$ for a pattern $X(\overline{y_n})$ in long $\beta\eta$-normal form. In the first-order case, a most general unifier is either empty or decreases the number of variables. For pattern unification, we also have the Flex-Flex Same rule with substitutions of the form $\{H \mapsto \lambda\overline{x_n}.H'(\overline{y_m})\}$, where $\{\overline{y_m}\} \subseteq \{\overline{x_n}\}$. Notice that such substitutions do not increase the size. In Section 5.1 we will show the following result, which is difficult to obtain with the rules of System PU: if θ is a most general unifier of two patterns s and t, then either $|\mathcal{FV}(\theta s)| < |\mathcal{FV}(s,t)|$, or θ is not size-increasing.

Deletion

$$[t =^? t \mid S] \quad \Rightarrow \quad S$$

Decomposition

$$[\lambda\overline{x_k}.f(\overline{t_n}) =^? \lambda\overline{x_k}.f(\overline{t'_n}) \mid S] \quad \Rightarrow \quad [\overline{\lambda\overline{x_k}.t_n =^? \lambda\overline{x_k}.t'_n} \mid S]$$

Elimination

$$[F =^? t \mid S] \quad \Rightarrow^{\theta} \quad \theta S \text{ if } F \notin \mathcal{FV}(t) \text{ and where } \theta = \{F \mapsto t\}$$

Imitation/Projection

$$[\lambda\overline{x_k}.F(\overline{y_n}) =^? \lambda\overline{x_k}.v(\overline{t'_m}) \mid S] \quad \Rightarrow^{\theta} \quad [\overline{\lambda\overline{x_k}.H_m(\overline{y_n}) =^? \lambda\overline{x_k}.\theta t'_m} \mid \theta S]$$

where $\theta = \{F \mapsto \lambda\overline{y_n}.v(\overline{H_m(\overline{y_n})})\}$, v is a constant or $v \in \{\overline{y_n}\}$, and $F \notin \mathcal{FV}(\lambda\overline{x_k}.v(\overline{t'_m}))$

Flex-Flex Same

$$[\lambda\overline{x_k}.F(\overline{y_n}) =^? \lambda\overline{x_k}.F(\overline{y'_n}) \mid S] \quad \Rightarrow^{\theta} \quad \theta S \text{ where } \theta = \{F \mapsto \lambda\overline{x_n}.F'(\overline{z_p})\}$$

and $\{\overline{z_p}\} = \{y_i \mid y_i = y'_i\}$

Flex-Flex Diff

$$[\lambda\overline{x_k}.F(\overline{y_n}) =^? \lambda\overline{x_k}.F'(\overline{y'_m}) \mid S] \quad \Rightarrow^{\theta} \quad \theta S$$

where $\theta = \{F \mapsto \lambda\overline{y_n}.H(\overline{z_p}),$
$F' \mapsto \lambda\overline{y'_m}.H(\overline{z_p})\}$
and $\{\overline{z_p}\} = \{\overline{y_n}\} \cap \{\overline{y'_m}\}$

Figure 4.3: System PU for Pattern Unification

Patterns have other important properties. A λ-term can be flattened to a pattern plus constraints as follows. For instance,

$$\lambda x.h(\lambda y.f(H(y,G(a))),G(X))$$

can be flattened to

$$\lambda x.h(\lambda y.f(X_1(x,y)),X_2)$$

with constraints

$$X_1 = \lambda x,y.H(y,G(a)), X_2 = G(X).$$

Formally, **flattening** a term t at position p yields the term $t[X(\overline{y_n})]_p$ with constraint $X = \lambda\overline{y_n}.t|_p$ where $\overline{y_n} = \mathcal{BV}(t,p)$ and X is new variable of appropriate type. Intuitively, the pattern part represents the rigid part of a term.

Proposition 4.2.3 *Assume p and q can be flattened to patterns p' and q' with the constraints C. If p' and q' do not unify then p and q do not unify either.*

4.3 Higher-Order Term Rewriting

We will in general follow the notation of first-order term rewriting, see e.g. [DJ90]. Our definitions for higher-order rewrite systems in this section are inspired from [MN97] and [Wol93]. We will often, but not in general require that the left-hand side be a higher-order pattern, as done in [Nip91a, Nip93c]. An important restriction is to use rules of base type only, as it simplifies the definition of the rewrite relation: it is close to the first-order case. There exist several similar notions of higher-order rewriting [Klo80, KOR93, OR94a, Pol94, Wol93]. The quite general approach in [Oos94, OR94b] separates higher-order rewriting into an rewriting part and a substitution part. The latter is just simply-typed λ-calculus in our case. Some of the earlier works [Klo80] use a different λ-calculus for substitution. For an overview we refer to [Oos94, OR94a].

Definition 4.3.1 A **rewrite rule** is a pair $l \rightarrow r$ such that l is not η-equivalent to a free variable, l and r are long $\beta\eta$-normal forms of the same base type, and $\mathcal{FV}(l) \supseteq \mathcal{FV}(r)$. A **General Higher-Order Rewrite System (GHRS)** is a set of rewrite rules.

Definition 4.3.2 Assuming a rule $(l \rightarrow r) \in R$ and a position p in a term s in long $\beta\eta$-normal form, a **rewrite step** from s to t is defined as

$$s \longrightarrow^{l \rightarrow r}_{p,\theta} t \;\Leftrightarrow\; s|_p = \theta l \wedge t = s[\theta r]_p.$$

We often omit some of the parameters $l \to r, p$ and σ of a rewrite step $\longrightarrow^{l\to r}_{p,\theta}$ and for a rewrite step with some rule from a GHRS R we write $s \longrightarrow^{R} t$.

Recall that we work with terms in long $\beta\eta$-normal form only, and consider this normalization as implicit, e.g. $l\theta = l\theta\updownarrow^{\eta}_{\beta}$. Notice that the subterm $s|_p$ may contain free variables which used to be bound in s. For instance $(\lambda x.f(g(x)))|_{1.1} = g(x)$. The following definition will be used to get a formal handle on these variables.

Definition 4.3.3 An $\overline{x_k}$**-lifter** of a term t **away from** W is a substitution $\sigma = \{F \mapsto (\rho F)(\overline{x_k}) \mid F \in \mathcal{FV}(t)\}$ where ρ is a renaming such that $\mathcal{D}om(\rho) = \mathcal{FV}(t)$, $\mathcal{R}ng(\rho) \cap W = \{\}$ and $\rho F : \tau_1 \to \cdots \to \tau_k \to \tau$ if $x_1 : \tau_1, \ldots, x_k : \tau_k$ and $F : \tau$.

For example, $\{G \mapsto G'(x)\}$ is an x-lifter of $g(G)$ away from any set of variables W not containing G'. For simplicity, we often assume that W contains all variables used so far and leave W implicit. A term t is $\overline{x_k}$**-lifted** if an $\overline{x_k}$-lifter has been applied to t. Similarly, a rewrite rule $l \to r$ is $\overline{x_k}$-lifted, if l and r are $\overline{x_k}$-lifted.

Now we can give an alternative definition for rewriting (see also [Fel92]). We have $s \longrightarrow^{l\to r}_{p,\theta} t$ if $\lambda\overline{x_k}.s|_p = \lambda\overline{x_k}.\theta l$ and $t = s[\theta r]_p$, where $\{\overline{x_k}\} = \mathcal{BV}(s,p)$ and $l \to r$ is $\overline{x_k}$-lifted away from $V \supseteq \mathcal{FV}(s)$.

For instance, consider the rewrite step

$$\lambda x.f(x) \longrightarrow^{f(Y)\to g(Y)}_{\{Y\mapsto x\}} \lambda x.g(x).$$

With the latter notion of rewriting, we first apply the lifter $\sigma = \{Y \mapsto Y'(x)\}$ to $f(Y) \to g(Y)$. Then $\lambda x.f(x) = \theta\lambda x.f(Y'(x))$ with $\theta = \{Y' \mapsto \lambda x.x\}$. For rewriting, lifting is unnecessary if loose bound variables are used, since only matching l with $s|_p$ is performed. However for narrowing, as developed later, unification is needed instead of matching and hence lifting is essential.

In contrast to the first-order notion of term rewriting, $\longrightarrow$ is not stable under substitution: reducibility of s does not imply reducibility of θs. For instance, if a is reducible then $F(a)$ is reducible as well, but $\{F \mapsto \lambda x.x\}F(a)$ is not. Its transitive reflexive closure is however stable:

Lemma 4.3.4 *Assume an GHRS R. If* $s \overset{*}{\longrightarrow}^{R} t$ *and* $\theta \overset{*}{\longrightarrow}^{R} \theta$ *then* $\theta s \overset{*}{\longrightarrow}^{R} \theta t$.

The proof of this seemingly simple lemma is rather involved and can be found in [MN97]; a similar result is shown in [LS93] for conditional rules.

A GHRS where all rules have patterns on the left-hand side is called **HRS**. This corresponds to the original definition in [Nip91a], which was renamed to PRS in [MN97]. We call a rule $l \to r$ **pattern rule,** if both l and r are patterns.

Furthermore, an HRS with pattern rules only is called a **pattern HRS**. A rule $l \to r$ is **left-linear**, if l is linear. An HRS is called left-linear, if it consists of left-linear rules.

For programming languages, the set of constants is often divided into **constructors** and defined symbols. A symbol f is called a **defined symbol**, if a rule $f(\ldots) \longrightarrow t$ exists. It is assumed that constructors are injective, i.e. $c(\overline{t}) = c(\overline{t'})$ iff $\overline{t} = \overline{t'}$, and that different constructors build different terms, i.e. $c(\overline{t}) \neq c'(\overline{t'})$ if $c \neq c'$. Constructor symbols are denoted by c and d. A term is a constructor term if no defined symbols occur.

We often identify an (G)HRS R with its associated rewrite relation. For instance, we say an (G)HRS R is terminating, if $\longrightarrow^R$ is terminating. A term is in R-**normal form** if no rule from R applies and a substitution θ is R-**normalized** if θX is in R-normal form for all $X \in \mathcal{D}om(\theta)$. For a term t we denote the R-normal form by $t\downarrow_R$, if uniquely defined, and similarly for substitutions.

A rewrite step $s \longrightarrow_p^{l \to r} t$ is **innermost** w.r.t. some GHRS R, if s is not R-reducible at a position below p. A sequence of reductions $s \stackrel{*}{\longrightarrow}^R t$ is innermost, if each step in the sequence is innermost. An **outermost** rewrite step is defined correspondingly as a step where no rewrite step applies above. In programming applications, innermost reduction corresponds to eager evaluation and outermost to call-by-name evaluation. We write $s \longrightarrow_{\neq \varepsilon}^{l \to r} t$ for a rewrite step that occurs below the root position of s.

Since β-reduction on patterns only renames bound variables, we obtain the following result on reducibility of substitutions. It generalizes the first-order case and is used for completeness of narrowing, as developed in Chapter 6.

Fact 4.3.5 *Assume an GHRS R, a pattern $\lambda \overline{y_k}.F(\overline{x_n})$, and a substitution θ. Then $\theta F(\overline{x_n})$ is R-reducible, iff θF is R-reducible.*

This result will be often used for higher-order patterns, where free variables occur only in the form as in the result above. However, this result does not hold if rules of non-base type are permitted. For instance, consider a non-base type rule $\lambda x.x \to \bot$. Then $\lambda x.x$ is reducible, but x is not.

This raises the question of how important the restriction to rules of base type is. The problematic case are rules with a binder outside, e.g. a rule

$$\lambda x.f(G(x),x) \to \lambda x.g(x)$$

The problem is that our notion of rewriting on terms in β-normal form is insufficient, as discussed in [Nip91a, MN97]: a term $f(g(a),a)$, which is equivalent to $(\lambda x.f(g(x),x))a)$ by β-reduction, cannot be rewritten with the above rule. Note that the above pattern rule is equationally equivalent (see next section for details) to the non-pattern rule

$$f(G(X),X) \to g(X)$$

which blurs the distinction between HRS and GHRS.

However, there are examples, as common in functional programming, where rules of functional type are convenient. This class of rules which is created by partial application can be modeled easily by supplying additional arguments. For instance, a rule

$$f(a,X) \to g(b,X)$$

can be abbreviated by

$$f(a) \to g(b)$$

This is equivalent, since we work on η-expanded terms: a term $f(a)$ can be rewritten by the first rule, since $f(a) = \lambda x.f(a,x)$.

4.3.1 Equational Logic

A rewrite system R induces an equivalence on terms. This equational theory $=_R$ is defined by the inference rules in Figure 4.4. In violation of our usual conventions, we do not assume for the rules in Figure 4.4 that the terms are in any normal form. It is shown in [Wol93] that the equivalence relation $=_R$ coincides with a particular model theoretic semantics for higher-order equational logic.

Notice that in the higher-order case the application rule implies the usual congruence rule of the form

$$\frac{t_1 =_R t'_1, \ldots, t_n =_R t'_n}{f(\overline{t_n}) =_R f(\overline{t'_n})}$$

Also, in the higher-order case, the standard substitution rule

$$\frac{s =_R t}{\theta s =_R \theta t}$$

can be inferred from the above by repeated abstractions and applications. For instance, assume $\theta = \{x \mapsto u\}$, then

$$\frac{\dfrac{s =_R t}{\lambda x.s =_R \lambda x.t} \quad \dfrac{}{u =_R u}}{(\lambda x.s)u =_R (\lambda x.t)u}$$

For higher-order equational theories, the following equivalence of the equational theory and term rewriting has first been shown for HRS in [Nip91a] and has been extended to GHRS in [MN97].

Rule	$\dfrac{}{l =_R r}$	$l \to r \in R$
Reflexivity	$\dfrac{}{t =_R t}$	
Symmetry	$\dfrac{s =_R t}{t =_R s}$	
Transitivity	$\dfrac{s =_R t \quad t =_R u}{s =_R u}$	
Abstraction	$\dfrac{s =_R t}{\lambda x.s =_R \lambda x.t}$	
Application	$\dfrac{s =_R s' \quad t =_R t'}{(s\, t) =_R (s'\, t')}$	
Conversion	$\dfrac{s =_{\alpha\beta\eta} t}{s =_R t}$	

Figure 4.4: Equational Theory of a GHRS R

Theorem 4.3.6 *For any GHRS R the following are equivalent:*

$$s =_R t \;\Leftrightarrow\; s\updownarrow_\beta^\eta \overset{*}{\leftrightarrow}_R t\updownarrow_\beta^\eta$$

The proof in [Wol93], which gives a similar result without restrictions on the left-hand sides, only holds for terms in β-normal form, as observed by Nipkow [MN97]. A similar result for conditional equations can be found in [LS93].

4.3.2 Confluence

Some of the important confluence criteria for first-order rewriting (see for instance [Klo92, Hue80]) have been lifted to the higher-order case. As in the first-order case, most confluence criteria are based on an analysis of overlaps:

Definition 4.3.7 A rule $l \to r$ of some HRS **overlaps** with a pattern t, if $\theta t \longrightarrow_p^{l\to r} s$ for some substitution θ at a non-variable position p in t. Since l and t are patterns, we assume that θ is the most general unifier of $t|_p$ and l (modulo lifting). Two rules $l_0 \to r_0$ and $l_1 \to r_1$ have an **overlap**, if $l_1 \to r_1$ overlaps with l_0 or vice versa.

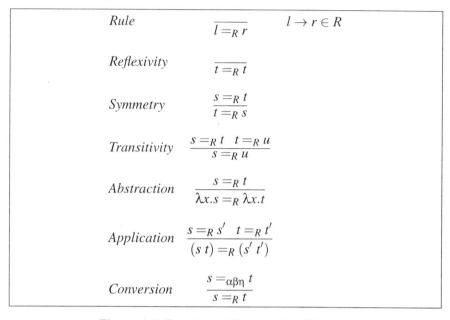

An HRS is called **orthogonal**, if it is left-linear and there are no overlaps. The following classical result of the first-order case has also been lifted to the higher-order case [MN97, Nip93c]:

Theorem 4.3.8 *Orthogonal HRS are confluent.*

For an overview of similar results in a general setting see [Oos94] (also in [OR94b]). Orthogonal HRS cover an important class of rewrite rules: (higher-order) functional programs are left-linear and either allow no overlaps, or only weak overlaps [Oos94], for which confluence holds as shown in [Oos94].

If there exist overlaps, they give rise to so-called critical pairs. A pair (u, v) is called a **critical pair** of $l_0 \to r_0$ and $l_1 \to r_1$ if the rules overlap at position p with substitution θ and $\theta l_i \stackrel{l_j \to r_j}{\longrightarrow}_p u$ and $v = \theta r_i$, where $i \in \{0,1\}$ and $j = 1 - i$. For instance, the rules $\mathit{diff}(\lambda y.G, X) \to 0$ and

$$\mathit{diff}(\lambda y.\mathit{sin}(F(y)), X) \to \mathit{cos}(F(X)) * \mathit{diff}(\lambda y.F(y), X)$$

overlap and give rise to the critical pair

$$(0, \mathit{cos}(G') * \mathit{diff}(\lambda y.G', X))$$

via the substitution $\{G \mapsto \mathit{sin}(G'), F \mapsto \lambda x.G'\}$. The well-known (first-order) critical pair lemma [KB70] also generalizes to HRS:

Theorem 4.3.9 ([MN97, Nip91a]) *An HRS R is locally confluent iff all critical pairs (u, v) are joinable, i.e. $u \downarrow_R v$.*

This yields the important result that confluence of terminating HRS is decidable, as local confluence implies confluence for terminating HRS. In the first-order case, the critical pair lemma also yields a method for constructing convergent rewrite rules, called completion [DJ90]. Its main idea, orienting critical pairs and adding them to the set of rewrite rules, fails in the higher-order case, since critical pairs may not be patterns.

For first-order rewriting, there is a difference between confluence and ground confluence [Höl89]. An HRS is **ground confluent** if it is confluent on ground terms of a fixed signature. For higher-order term rewriting ground confluence and confluence coincide, as ground terms may contain local "constants" in the form of bound variables.[1] Then, as in the first-order case without the restriction to a certain signature both are equivalent. (See [Höl89] for a detailed discussion.)

---

[1] More formally, assume a rewrite system is ground confluent but not confluent. Then for any non-joinable pair of terms s, t, there are ground terms $\lambda\overline{x_n}.s, \lambda\overline{x_n}.t$ which are joinable iff s, t are, which is a contradiction. The other direction is trivial.

4.3.3 Termination

Termination of rewriting is undecidable, but there exist many results for terminating classes of rewrite systems or semi-decision procedures (see for instance [DJ90]). A term ordering $<$ is called a **termination ordering** of some HRS R if $\longrightarrow^R \subseteq >$ and $<$ terminates. Usually, to show termination for an HRS $R = \{\overline{l_n \to r_n}\}$, one has to find an ordering $<$ with $\overline{l_n > r_n}$ that extends to $\longrightarrow^R$. For the first-order case, there exist large classes of orderings that are known to extend to $\longrightarrow^R$. In the higher-order case, such orderings are more difficult, as $>$ must be preserved by higher-order substitutions.

One current approach [Kah95, Pol94] aims at finding appropriate semantic domains in order to interpret higher-order rewriting. In particular, strictly monotonic interpretations of terms in monotonic domains are used in [Pol94, PS95]. That is, (higher-order) symbols are interpreted by monotonic functions. It is shown that an HRS terminates if the interpretation of the right hand-side is smaller than the left-hand side for each rule. Alternatives which extend syntactic orderings on first-order terms [DJ90] to higher-order terms are pursued in [LS93, ALS94a, LP95, JR96]. These papers use first-order interpretations on (often restricted) λ-terms. A main problem is that basic first-order properties of orderings, e.g. stability under substitutions, are more involved in λ-calculus.

A confluent and terminating HRS R is called **convergent**. It follows from Theorem 4.3.6 for convergent R that $s =_R t$ can be decided by comparing $s{\downarrow_R}$ and $t{\downarrow_R}$. For a terminating GHRS R, we define $\longrightarrow^R_{sub}$ as

$$\longrightarrow^R_{sub} = \longrightarrow^R \cup >_{sub} .$$

For the first-order case, termination of this reduction was shown in [JK86]. The proof in the latter can be generalized to the higher-order case as follows. We first need the following trivial lemmata.

Lemma 4.3.10 *If $s \longrightarrow^R t$ for a GHRS R and s is a subterm of s', i.e. $s'|_p = s$, then $s' \longrightarrow^R s'[t]_p$.*

Lemma 4.3.11 *Assume an GHRS R. If*

$$s >^*_{sub} t \longrightarrow^R u$$

then there exists $t' >_{sub} t$ such that

$$s \longrightarrow^R t' >^*_{sub} u.$$

Theorem 4.3.12 *The reduction* $\longrightarrow^R_{sub} = \longrightarrow^R \cup >_{sub}$ *is terminating for a GHRS R if* $\longrightarrow^R$ *is terminating.*

Proof by contradiction. Assume an infinite sequence of $\longrightarrow^R_{sub}$ reductions. If the reduction does not contain some $>_{sub}$-step or only $>_{sub}$-steps, we clearly have a contradiction. Otherwise, assume the first $>_{sub}$-step occurs after a sequence of n $\longrightarrow^R$-steps. Then by Lemma 4.3.11, we can construct a sequence of length $n+1$. Repeating this yields a contradiction. □

Chapter 5

Decidability of Higher-Order Unification

As the known decidability results (see Section 2.6) do not cover many practical cases, we examine decidability of higher-order unification more closely, mostly considering the second-order case. An overview of the results and the corresponding sections can be found in Figure 5.1. Notice that the results only hold under the assumption that all conditions in the path to the node hold.

We show in Section 5.2.1 that unification of a linear higher-order pattern with an arbitrary second-order term is decidable and finitary, if the two terms share no variables. In particular, we do not have to resort to pre-unification, as equations with variables as outermost symbols on both sides (flex-flex) pairs can be finitely solved in this case. Further extensions are discussed in Section 5.2.2. For instance, unifying two second-order terms, where one term is linear, is shown to be undecidable if the terms contain bound variables and decidable otherwise. Then we develop an extension of higher-order patterns with decidable unification in Section 5.3, where second-order linear variables are permitted. The case with repeated variables is discussed in Section 5.3.2. The main result here is that unification of "induction schemes", e.g. $\forall x.P(x) \Rightarrow P(x+1)$, with first-order terms is decidable.

5.1 Elimination Problems

In this section we consider a particular class of unification problems, called **elimination problems**, of the form

$$\lambda \overline{x_n}.P(\overline{y_m}) =^? \lambda \overline{x_n}.t,$$

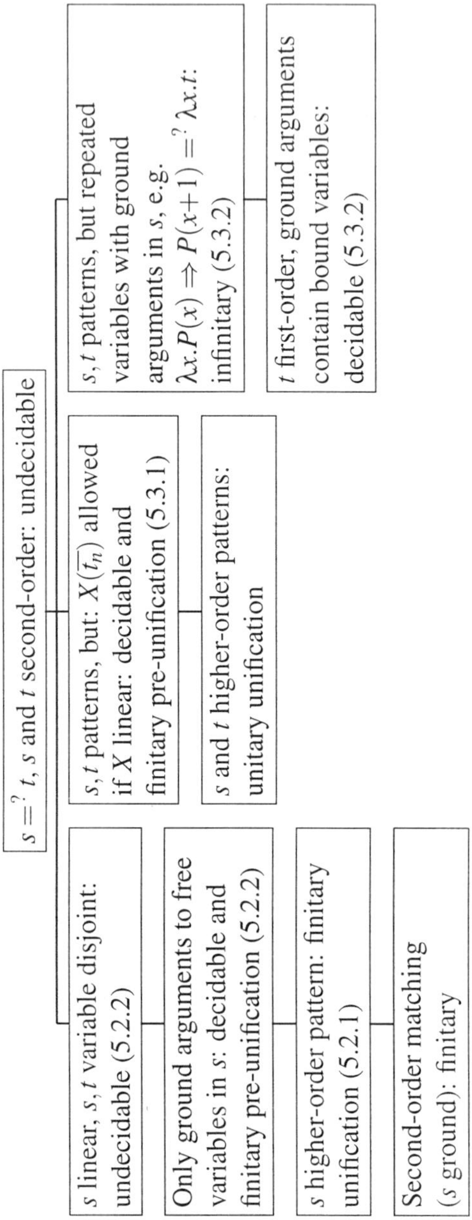

Figure 5.1: Results on Second-Order Unification

where $P \notin \mathcal{FV}(t)$. In the first-order case such equations are trivially solvable, here such an equation may not have a solution due to bound variables. For instance, the unification problem $\lambda x, y.P(x) =^? \lambda x, y.f(y)$ has no solution. Among the applications of elimination problems are certain flex-flex pairs. This will allow us later to use unification instead of pre-unification in some cases.

We call this class elimination problems, as they generalize first-order elimination. Secondly, the strategy to solve such goals is to eliminate the bound variables $\{\overline{x_n}\} - \{\overline{y_m}\}$ in t by appropriate substitutions. For instance, the equation

$$\lambda x, y.P(x) =^? \lambda x, y.f(x, X(y))$$

has the most general solution $\{X \mapsto \lambda z.X', P \mapsto \lambda x.f(x, X')\}$. This example actually falls into the class of patterns and is thus solvable by System PU. The main difference is that System PU introduces many temporary variables for partial bindings for P. Intuitively, all we need for solving $\lambda \overline{x_n}.P(\overline{y_m}) =^? \lambda \overline{x_n}.t$, where t is a pattern, is the following:

- Let $W = \{\overline{x_n}\} - \{\overline{y_m}\}$.
- If some $x \in W$ occurs on a rigid path in t then fail. Otherwise,
- for each occurrence of free a variable $X(\overline{z_n})$ in t, bind X to a term $\lambda \overline{z_n}.X'(\{\overline{z_n}\} \cap W)$, where X' is a new variable of appropriate type.

Hereby the last expression assumes an arbitrary conversion of the set of arguments to X' to a list. The reason why we explain this special case in such detail is that this strategy is actually used in implementations of PU, see e.g. [Nip93b].

In addition, this strategy shows that for solving elimination problems, no "real" new variables have to be introduced, only the variables in t are mapped to new variables with fewer arguments. As in addition P is bound to some term, the total number of variables decreases, which will be important for some results in Section 5.3.

The main focus of this section is on elimination problems where t is an arbitrary second-order term. For this case, there can be many different solutions to an elimination problem, as the next example shows:

Example 5.1.1 Consider the pair

$$\lambda x, y.F(x) =^? \lambda x, y.F'(F''(x), F''(y)).$$

There are two ways to eliminate y on the rhs, i.e. $\theta_1 = \{F' \mapsto \lambda z_1, z_2.F'_1(z_1)\}$ and $\theta_2 = \{F'' \mapsto \lambda z_1.F''_1\}$, where F'_1 and F''_1 are new variables.

Eliminate

$$(\theta, [\lambda\overline{x_k}.P(\overline{t_n})|R], W) \Rightarrow_{EL} (\tau_{P,i}\theta, \tau_{P,i}[\lambda\overline{x_k}.P(\overline{t_n})|R], W) \quad \text{if } \exists x \in W \cap \mathcal{BV}(\lambda\overline{x_k}.t_i)$$

Proceed

$$(\theta, [\lambda\overline{x_k}.v(\overline{t_n})|R], W) \Rightarrow_{EL} (\theta, [\overline{\lambda\overline{x_k}.t_n}|R], W), \quad \text{unless the atom } v \text{ is a bound variable in } W$$

Figure 5.2: System EL for Eliminating Bound Variables

We first need some notation to formalize these ideas. For a variable F of type $\overline{\alpha_n} \to \alpha_0$ we define the **i-th parameter eliminating substitution** $\tau_{F,i}$ as

$$\tau_{F,i} = \{F \mapsto \lambda\overline{x_n}.F'(x_1, \ldots, x_{i-1}, x_{i+1}, \ldots, x_n)\},$$

where F' is a new variable of appropriate type.

The transformation rules $\Rightarrow_{EL}$ in Figure 5.2 transform triples of the form (θ, l, W), where θ is the computed substitution, l is the list of remaining terms, and W is the set of bound variables to be eliminated. We say system EL succeeds if it reduces a triple to $(\theta, [], W)$. For the flex-flex pair in Example 5.1.1 system EL works as follows, starting with the triple

$$(\{\}, [\lambda x, y.F'(F''(x), F''(y))], \{y\}).$$

EL can either eliminate the second argument of F' or it can proceed until the triple $(\{\}, [\lambda x, y.F''(y)], \{y\})$ is reached and then eliminate y. In these two cases, EL succeeds with θ_1 and θ_2, respectively, as in Example 5.1.1. All other cases fail.

Observe that system EL is not optimal, as it can produce the same solution twice. For instance, consider the pair $\lambda x.F =^? \lambda x.F'(F'(x))$. There are two different transformation sequences that yield the unifier $\{F' \mapsto \lambda s.F'', \ldots\}$. More precisely, this happens only if a bound variable occurs below nested occurrences of a variable at subtrees with the same index.

We first show the correctness of EL.

Lemma 5.1.2 (Correctness of EL) *Let $\lambda\overline{x_k}.P(\overline{y_m}) =^? \lambda\overline{x_k}.t$ be a pair where P does not occur in t. Assume further $W = \{\overline{x_k}\} - \{\overline{y_m}\}$. If $(\{\}, [t], W) \overset{*}{\Rightarrow}_{EL} (\theta, [], W)$ then $\theta \cup \{P \mapsto \theta\lambda\overline{y_m}.t\}$ is a unifier of $\lambda\overline{x_k}.P(\overline{y_m}) =^? \lambda\overline{x_k}.t$.*

Proof We show that $\{P \mapsto \theta\lambda\overline{y_m}.t\}$ is a well-formed substitution, i.e. all bound variables in θt are locally bound or are in $\overline{y_m}$. As any successful sequence of EL reductions must traverse the whole term $\lambda\overline{x_k}.t$ to succeed, only bound variables in $\{\overline{y_m}\}$ can remain; occurrences of $\{\overline{x_k}\} - \{\overline{y_m}\}$ are either eliminated by some substitution $\tau_{P,i}$ in rule Eliminate, or the algorithm fails as the rule Proceed does not permit these bound variables. □

The next lemma states that if θ eliminates all occurrences of variables in W from $\overline{t_n}$, then there is a sequence of EL reductions that approximates θ.

Lemma 5.1.3 *If* $\tau[\overline{t_n}] = [\overline{t_n}]$, $\mathcal{BV}(\overline{\theta t_n}) \cap W = \emptyset$, $\theta = \delta\tau$ *for some substitution* δ, *and* $\overline{t_n}$ *are weakly second-order terms, then there exist a reduction* $(\tau, [\overline{t_n}], W) \stackrel{*}{\Rightarrow}_{EL} (\theta', [], W)$ *and a substitution* δ' *such that* $\theta = \delta'\theta'$.

Proof by induction on the sum of the sizes of the terms in $[\overline{t_n}]$. Clearly, each $\Rightarrow_{EL}$ reduction reduces this sum. The base case, where $n = 0$, is trivial. We show that for each such problem some EL step applies and that the induction hypothesis can be applied. Depending on the form of t_1 and the conditions of the rules of EL, we apply different rules. Assume t_1 is of the form $\lambda\overline{x_k}.P(\overline{u_m})$ and $\theta P = \lambda\overline{y_m}.t$. By our variable conventions, we can assume that $W \cap \mathcal{BV}(\theta P) = \emptyset$. As $\lambda\overline{x_k}.P(\overline{u_m})$ is a weakly second-order term, some bound variable from W appears in $\theta\lambda\overline{x_k}.P(\overline{u_m})$ if and only if it appears in some $\theta\lambda\overline{x_k}.u_i$ where $y_i \in \mathcal{BV}(\lambda\overline{y_m}.t) = \mathcal{BV}(\theta P)$: if some u_k is a bound variable, then only renaming takes place, otherwise, u_k must be first-order and hence y_k must occur at a leaf in t. Then let

$$i = Min\{j \mid \exists x \in \mathcal{BV}(\theta\lambda\overline{x_k}.u_j) \cap W\}.$$

The above set describes the indices of bound variables that may not occur in $\theta P = \lambda\overline{y_m}.t$ by assumption on θ, e.g. $y_i \notin \mathcal{BV}(\lambda\overline{y_m}.t)$. If the above set is empty and no j exists, we apply the second rule and can then safely apply the induction hypothesis.

In case the minimum i exists, we have $\mathcal{BV}(\theta\lambda\overline{x_k}.u_i) \cap W \subseteq \mathcal{BV}(\lambda\overline{x_k}.u_i) \cap W$. Hence the Eliminate rule applies with

$$\tau_{P,i} = \{P \mapsto \lambda\overline{x_m}.P_0(x_1,\ldots,x_{i-1},x_{i+1},\ldots,x_m)\}.$$

Then we can apply the induction hypothesis to $(\tau_{P,i}\tau, \tau_{P,i}[\overline{t_n}], W)$: define δ' such that $\delta' X = \delta X$ if $X \neq P$ and $\delta' P_0 = \lambda y_1,\ldots,y_{i-1},y_{i+1},\ldots,y_m.t$. Notice that δ' is well-formed, as $y_i \notin \mathcal{BV}(\lambda\overline{y_m}.t)$. Clearly, the premises for the induction hypothesis are fulfilled, as $\theta = \delta'\tau_{P,i}\tau$ follows from $\tau P = P$. Then the induction hypothesis assures that both EL succeeds with a substitution θ' and that a substitution δ'' exists such that $\theta = \delta''\theta'$.

The remaining cases of t_1 are trivial as the Proceed rule does not compute substitutions. □

Now we can show that EL captures all unifiers. We use EL to solve elimination problems of the form $\lambda\overline{x_k}.P(\overline{y_m}) =^? t$, where t is not η-equivalent to a free variable. In the latter case the solution considered in the next lemma introduces more new variables than the trivial solution $t \mapsto \lambda\overline{x_k}.P(\overline{y_m})$.

Lemma 5.1.4 (Completeness of EL) *Assume θ is a unifier of a pair of the form $\lambda\overline{x_k}.P(\overline{y_m}) =^? \lambda\overline{x_k}.t$, where $\lambda\overline{x_k}.t$ is not η-equivalent to a free variable and $\lambda\overline{x_k}.P(\overline{y_m})$ is a pattern. Assume further $\lambda\overline{x_k}.t$ is weakly second-order and does not contain P. Let $W = \{\overline{x_k}\} - \{\overline{y_m}\}$. Then there exist a substitution $\theta'' = \theta' \cup \{P \mapsto \theta'\lambda\overline{y_m}.t\}$ and a reduction $(\{\}, [\lambda\overline{x_k}.t], W) \overset{*}{\Rightarrow}_{EL} (\theta', [], W)$ such that θ'' is more general than θ.*

Proof It is clear that any unifier must eliminate all bound variables from W on the right-hand side. Then the proof follows easily from Lemma 5.1.3. □

It can be shown that EL computes at most a quadratic number of different substitutions. Let n be the number of occurrences of variables to be eliminated and let m be the maximal number of nested free variables. Then there can be at most m distinct ways to eliminate some particular variable. As m and n are both linear in the size, the maximal number of solutions, i.e. mn, is quadratic.

Observe that EL is not complete for the third-order case. Here, if a free variable has two arguments, one can be a function. If in some solution this function is applied to the other argument, then this function could eliminate, in the above sense, the other argument. For instance, consider the third-order pair $\lambda x,y.F(x) =^? \lambda x,y.F'(\lambda z.F''(z),y)$. Here EL would not uncover the solution

$$\{F' \mapsto \lambda y,z.F'_0(y(z)), F'' \mapsto \lambda x.a, F \mapsto \lambda y,z.F'_0(a)\}.$$

With System EL we can show the following result on pattern unification much easier than with System PU, as EL introduces fewer variables.

Lemma 5.1.5 *Assume θ is a most general unifier of two patterns s and t, then either $|\mathcal{FV}(\theta s)| < |\mathcal{FV}(s,t)|$, or θ is not size-increasing.*

Proof Assume a reduction $[s =^? t] \overset{*}{\Rightarrow}{}^{\theta}_{PU} \overline{G_n}$. If no Elimination, Imitation or Projection is applied, then the substitution is not size increasing; this is trivial for Deletion and Decomposition and simple for the Flex-Flex rules. Otherwise, we apply System PU, but use EL instead for all equations of the form $\lambda\overline{x_n}.F(\overline{y_m}) =^? \lambda\overline{x_n}.t$. For this case, we show that a solution to such an equation reduces the number of variables. It is evident that the number of free variables remains unchanged under the parameter eliminating substitution δ computed by EL. As F is bound to some term $\lambda\overline{x_n}.\delta t$, the number of variables reduces.

□

It would be interesting to develop deterministic and efficient implementations of EL that compute the set of all unifiers. For instance, if a variable from W occurs on a path where no free variable occurs, then this branch can safely fail. Furthermore, an effective version should also detect when it produces the same solution twice.

Repeated Bound Variables

In the last section, we did not allow repeated bound variables on the left-hand side. In the next lemma we extend this result to relaxed patterns, which causes some technical overhead. Repeated variables may cause an additional number of distinct unifiers in each case, as there can be different permutations if a repeated variable occurs in the common instance. Consider for example the pair $\lambda x.F(x,x) =^? \lambda x.c(x)$. There are the two solutions $\{F \mapsto \lambda y,z.c(y)\}$ and $\{F \mapsto \lambda y,z.c(z)\}$.

As evident from this example, there can be an exponential number of incomparable unifiers in the general case. Consider for instance $\lambda x.F(x,x) =^? \lambda x.v$, where x occurs in v exactly n times. Then there are 2^n different solutions. Although this may seem very impractical, we conjecture that large numbers of unifiers are rare.

In the following result we do not formalize these possible permutations explicitly. For simplicity, we only specify the properties of the correct permutations. As the number of permutations is clearly finite, this is sufficient, but does not yield an effective algorithm for computing these.

Lemma 5.1.6 *A unification problem $\lambda\overline{x_k}.P(\overline{y_m}) =^? t$ where $\lambda\overline{x_k}.P(\overline{y_m})$ is a relaxed pattern and t is weakly second-order and does not contain P, is finitely solvable.*

Proof Consider a pair $\lambda\overline{x_k}.P(\overline{y_m}) =^? \lambda\overline{x_k}.t$ and assume some bound variables occur several times in $P(\overline{y_m})$. Assume EL succeeds with $(\theta, [], \{\overline{x_k} - \overline{y_m}\})$. Let $p(i,j)$ be the position of the j-th occurrence of x_i in θt. For this solution of EL, all solutions for P are of the form $\{P \mapsto \lambda\overline{z_m}.t'\}$, where $Head(t'|_{p(i,j)}) = z_i$ and $y_i = x_i$ for all positions $p(i,j)$ of some x_i in θt and $Head(t'|_q) = Head(\theta t|_q)$ otherwise. Here the last equations allow for many permutations, as some x_j may occur repeatedly in $\overline{y_m}$. All these permutations are clearly independent from the remaining parts of the computed unifier, as P does not occur elsewhere, and can easily be computed. □

5.2 Unification of Second-Order with Linear Terms

As second-order unification is undecidable, we are interested in identifying decidable subclasses. The restriction discussed here is that one term of the unification problem is linear, i.e. has no repeated variables. We present in the following several results on the decidability of such unification problems, which range from finitary unification over finitary pre-unification to pure decidability. A major application of the results is narrowing with left-linear rules, as discussed in Section 6.3. In Section 6.3.1 we will extend the results in this section to sets of equational goals.

5.2.1 Unifying Linear Patterns with Second-Order Terms

In this section we show that unification of second-order λ-terms with linear patterns is decidable and finitary. We first use System PT to solve the pre-unification problem, followed by EL for solving the remainder. We use in the following weakly second-order terms, which is used in the next chapter.[1]

Lemma 5.2.1 *System PT terminates for two variable-disjoint terms $s =^? t$ if s is a linear pattern and t is weakly second-order. Furthermore, PT terminates with a set of flex-flex pairs of the form $\lambda\overline{x_k}.P(\overline{y_i}) =^? \lambda\overline{x_k}.P'(\overline{u_i})$ where all y_i are bound variables and P is isolated.*

Proof We show that system PT terminates for this unification problem. We start with the goal $s =^? t$ and apply the transformations modulo commutativity of $=^?$ in Figure 4.1. By this we achieve that after any sequence of transformations, all free variables on the left-hand sides (lhs) are isolated in the system of equations, as all newly introduced variables on the lhs are linear also. The latter can easily be seen by examining the cases for Imitation and Projection, the other rules are trivial. Another important invariant is that the left-hand sides remain patterns, which is easy to verify.

Since the first three transformations preserve the set of solutions, as shown in [SG89], we assume that Decomposition is applied after applying Projection to a lhs. We do not apply Elimination to flex-flex pairs, which could increase the size of some right-hand side (rhs) if a bound variable occurs repeatedly on the lhs, e.g. $\lambda x.c(x,x) =^? G$.

We use the following lexicographic termination ordering on the multiset of equations:

A: Compare the number of constant symbols on all lhs's, if equal

[1] This extends the earlier results in [Pre94a].

B: compare the number of occurrences of bound variables on all lhs's that are not below a free variable, if equal

C: compare the multiset of the sizes of the right-hand sides.

Now we show that the transformations reduce the above ordering:

Deletion trivial

Decomposition A or B is reduced.

Elimination Although this transformation eliminates one equation, it is not trivial that it also reduces the above ordering. Consider the possible equations Elimination is applied to:

- $F =^? \lambda\overline{x_k}.t$: as the free variable F is isolated, A and B remain constant and C is reduced.
- $\lambda\overline{x_k}.a(\ldots) =^? F$: the elimination of an equation with a constant a reduces A.
- $\lambda\overline{x_k}.x_i(\ldots) =^? F$: here B is reduced (and possibly A).

Imitation We have two cases:

- $\lambda\overline{x_k}.F(\overline{y_n}) =^? \lambda\overline{x_k}.f(\overline{t_m})$: the imitation binding for F is of the form $F \mapsto \lambda\overline{x_n}.f(\overline{H_m(\overline{x_n})})$. Now, we replace the above equation by a set of equations of the form $\lambda\overline{x_j}.H_i(\overline{y_n}) =^? \lambda\overline{x_j}.t_i$, where $i = 1,\ldots,m$. Notice that the number of constants on the lhs (A) does not increase, as all y_m are bound variables. Also, B remains unchanged. As F is isolated and hence does not occur on any right-hand side, C decreases.
- $\lambda\overline{x_k}.f(\overline{t_n}) =^? \lambda\overline{x_k}.F(\overline{u_m})$: we obtain an imitation binding as above. Then the number of constant symbols on the lhs's decreases, since F may not occur on the lhs's.

Projection We again have two cases:

- $\lambda\overline{x_k}.F(\overline{y_n}) =^? \lambda\overline{x_k}.y_i(\overline{t_m})$: as $\overline{y_n}$ are bound variables, this rule applies only if the head of the rhs is a bound variable as well, say y_i. Then the case is similar to the Imitation case above, as after Projection, the Decomposition rule applies.

- $\lambda\overline{x_k}.v(\overline{t_n}) =^? \lambda\overline{x_k}.F(\overline{u_m})$: as we have weakly second-order variables on the rhs, we again have two cases. If v is a bound variable, Decomposition applies after Projection and we proceed as in the Imitation case. In the remaining cases, projection bindings are of the form $F \mapsto \lambda\overline{x_m}.x_i$, where x_i is first-order. Hence the lhs's (i.e. A and B) are unchanged, whereas C decreases, as we assume terms in long $\beta\eta$-normal form.

□

So far, we have shown that pre-unification is decidable. To solve the remaining flex-flex pairs, notice that all of these are elimination problems of the form

$$\lambda\overline{x_k}.P(\overline{y_m}) =^? \lambda\overline{x_k}.P'(\overline{u_n}),$$

where P is isolated and $\{\overline{y_m}\}$ are bound variables.

Theorem 5.2.2 *Assume t is a weakly second-order λ-term and s is a linear pattern such that s shares no variables with t. Then the unification problem $s =^? t$ is decidable and finitary.*

Proof From Lemma 5.2.1 we know that PT terminates with a set of flex-flex pairs, where the lhs is a pattern. Then by Lemma 5.1.6 we can use EL to compute a complete and finite set of unifiers for some flex-flex pair, as EL terminates and is finitely branching. This unifier is applied to the remaining equations. Repeat this for all flex-flex pairs. This procedure terminates and works correctly as all lhs's are patterns and only have isolated variables. Notice that a flex-flex pair remains flex-flex when applying a unifier computed by EL. □

We have shown in Section 5.1 that EL may compute an exponential number of solutions when repeated variables are permitted. Clearly, the most concise representation of all unifiers is still a flex-flex pair. Which representation is best clearly depends on the application. For instance, flex-flex pairs may not be satisfactory for programming languages where explicit solutions are desired. For automated theorem proving, flex-flex pairs are a more compact representation and may reduce the search space.

It should also be noted that the unification problem in Lemma 5.2.1 allows for some nice optimizations for implementors. For instance, no occurs check is needed: the proof of Lemma 5.2.1 uses the invariant that all variables on the left-hand sides are isolated. Hence no variable can occur on a left-hand side and at the same time on some right-hand side.

5.2.2 Extensions

In the following sections, we will examine extensions of the above decidability result. First, notice that the linearity restriction is essential; otherwise full second-order unification can easily be embedded. But even with one linear term, this embedding still works:

Example 5.2.3 Consider the unification problem

$$\lambda x.F(f(x,G)) =^? \lambda x.g(f(x,t_1),f(x,t_2)),$$

where t_1 and t_2 are arbitrary second-order terms. By applying the transformations PT it is easy to see (compare to Example 4.1.3) that in all solutions of the above problem $F \mapsto \lambda x.g(x,x)$ and $t_1 =^? t_2$ must be solved, which is clearly undecidable.

Notice that this example requires a function symbol of arity two whereas second-order unification with monadic function symbols is decidable.

Motivated by this example, we consider the following two extensions. First, we assume that arguments of free variables are second-order ground terms. Secondly, we consider the case where an argument of a free variable contains no bound variables. These two cases can be combined in a straightforward way, as shown towards the end of this section. Thus arguments of free variables may either be ground second-order terms or terms with no bound variables. The generalization where only one term is linear follows easily from Example 5.2.3:

Corollary 5.2.4 *It is undecidable to determine if two second-order terms unify, even if one is linear.*

Pre-unification of two linear second-order terms without bound variables is however decidable and finitary, as shown by Dowek [Dow93]. This result is generalized in Section 5.3 to higher-order patterns with linear second-order variables.

Ground Second-Order Arguments to Free Variables

We now loosen the restriction that one term must be a linear pattern. As long as all arguments of free variables are either bound variables or ground second-order terms, we can still solve the pre-unification problem. In particular, for the second-order case, this can be rephrased as disallowing nested free variables. However, we only solve the pre-unification problem, as the resulting flex-flex pairs are more intricate than in the last section.

Similar to the above, we present a termination ordering for a particular strategy of the PT transformations. We will see that in essence only one new case results from these ground second-order terms. This case can be handled separately by second-order matching, which is decidable and finitary. (It is also an instance of Theorem 5.2.2.) That is, whenever such a matching problem occurs, this is solved immediately (considering all its solutions). Hence we first need a lemma about matching.

Lemma 5.2.5 *Solving a second-order matching problem with System PT only yields solutions that are ground substitutions.*

Proof by induction on the length of the transformation sequence. The base case, length zero, is trivial. The induction step has the following cases:

Deletion, Decomposition trivial

Elimination Consider the equation to which Elimination is applied:

$$\lambda\overline{x_k}.t =^? F$$

The claim is trivial as $\lambda\overline{x_k}.t$ is ground.

Imitation

$$\lambda\overline{x_k}.a(\overline{t_n}) =^? \lambda\overline{x_k}.F(\overline{u_m})$$

The imitation binding for F is of the form $F \mapsto \lambda\overline{y_m}.a(\overline{H_n(\overline{y_m})})$. Now, we replace the above equation by a set of equations of the form

$$\lambda\overline{x_k}.t_i =^? \lambda\overline{x_k}.H_i(\overline{u_m})$$

Clearly, for any matcher θ, $H_i \in \mathcal{D}om(\theta)$, and by induction hypothesis θH_i is ground. Hence in the solution to $\lambda\overline{x_k}.a(\overline{t_n}) =^? \lambda\overline{x_k}.F(\overline{u_m})$, F is mapped to a ground term.

Projection As we have second-order variables, we only have projection bindings of the form $F \mapsto \lambda\overline{y_m}.y_i$, which are trivially ground.

□

This result does not hold for the higher-order case, as observed by Dowek [Dow93]: e.g. $\{F \mapsto \lambda x.x(Y)\}$ is a solution to $F(\lambda x.a) =^? a$, but no complete set of ground matchers exists. Now we can show the desired theorem:

Theorem 5.2.6 *Assume s,t are λ-terms such that t is second-order, s is linear and s shares no variables with t. Furthermore, all arguments of free variables in s are either*

- *bound variables of arbitrary type or*
- *second-order ground terms of base type.*

Then the pre-unification problem $s =^? t$ is decidable and finitary.

Proof We give a termination ordering for system PT with the same additional assumptions as in the proof of Lemma 5.2.1. In addition, we consider solving a second-order matching problem as an atomic operation, with possibly many solutions. In particular, after a projection on a lhs, this step eliminates one equation and applies a (ground) substitution to the rhs. It is easy to see that the two premises, only isolated variables and no nested free variables on the lhs's, are invariant under the transformations.

We use the following (lexicographic) termination ordering on the multiset of equations:

A: Compare the number of occurrences of constant symbols and of bound variables that are not below a free variable on a lhs, if equal

B: compare the number of free variables in all rhs's, if equal

C: compare the multiset of the sizes of the rhs's.

Now we show that the transformations reduce the above ordering:

Deletion trivial

Decomposition A is reduced.

Elimination Although one equation is eliminated, it is not trivial that it also reduces the above ordering. Consider the equations this rule is applied to:

- $F =^? \lambda\overline{x_k}.t$: as the free variable F is isolated, A and B remain constant and C is reduced.
- $\lambda\overline{x_k}.v(\ldots) =^? F$: the elimination of an equation with a constant or bound variable v reduces A, as F does not occur on any rhs.

Imitation We have two cases, where a is a constant:

- $\lambda\overline{x_k}.F(\overline{u_m}) =^? \lambda\overline{x_k}.a(\overline{t_n})$: the imitation binding for F is of the form $F \mapsto \lambda\overline{y_m}.a(\overline{H_n(\overline{y_m})})$. Now we replace the above equation by a set of equations of the form $\lambda\overline{x_j}.H_i(\overline{u_m}) =^? \lambda\overline{x_j}.t_i$. Notice that the number of constants and bound variables not below a free variable on the lhs's (A) does not increase. As F is an isolated variable and does not occur on any right-hand side, B remains unchanged and C decreases.

- $\lambda\overline{x_k}.a(\overline{t_n}) =^? \lambda\overline{x_k}.F(\overline{u_m})$: we obtain an imitation binding as above, and the number of constant symbols on the lhs's (i.e. A) decreases, since F may not occur on the lhs.

Projection We again have two cases:

- $\lambda\overline{x_k}.F(\overline{t_m}) =^? \lambda\overline{x_k}.v(\overline{u_k})$: since F is an isolated variable, we obtain a single matching problem $\lambda\overline{x_k}.t_i =^? \lambda\overline{x_k}.v(\overline{u_k})$ or, if the i-th argument is a bound variable, the proof works as the case above (similar to the proof of Theorem 5.2.2). In the former case, any solution to this is a ground substitution by Lemma 5.2.5. Hence either B is reduced or, if the substitution is empty, B remains unchanged and C decreases.
- $\lambda\overline{x_k}.v(\overline{t_n}) =^? \lambda\overline{x_k}.F(\overline{u_m})$: as we have second-order variables on the rhs, we only have projection bindings of the form $F \mapsto \lambda\overline{y_m}.y_i$. Then the lhs's (i.e. A) are unchanged and both B and C decrease.

□

It might seem tempting to apply the same technique to arguments that are third-order ground terms, as third-order matching is known to be decidable. However, there can be an infinite number of matchers and without a concise representation for these the extension of the above method seems difficult.

No Bound Variables in an Argument of a Free Variable

We show that the remaining case, where an argument of a free variable contains no (outside-)bound variables, can be reduced to a simpler case. This method checks unifiability, but does not give a complete set of unifiers. The set of all outside bound variables occurring in a term $t = \lambda\overline{x_n}.v(\ldots)$ is written as $\mathcal{OBV}(t) = \{\overline{x_n} \cap \mathcal{BV}(t)\}$.

Theorem 5.2.7 *Assume $s = u[H(t_1,\ldots,t_i,\ldots,t_n)]_p$ and t are variable disjoint λ-terms such that s is linear. Assume further $\mathcal{OBV}(\lambda\overline{y_m}.t_i) = \emptyset$, where $\overline{y_m} = \mathcal{BV}(s,p)$. Then the unification problem $s =^? t$ has a solution, iff*

$$\lambda x_0.u[H(t_1,\ldots,x_0,\ldots)] =^? \lambda x_0.t,$$

where x_0 does not occur elsewhere, is solvable.

Proof Consider the unification problem

$$u[H(t_1,\ldots,t_i,\ldots,t_n)]_p =^? t$$

where H occurs only once in $u[H(t_1,\ldots,t_i,\ldots)]_p$ and t_i does not contain bound variables. Assume $\{X_1,\ldots,X_m\} = \mathcal{FV}(t_i)$. Let a solution to this problem be of the form $\{H \mapsto \lambda\overline{x_n}.t_0\} \cup \{\overline{X_o \mapsto u_o}\} \cup S$. As H does not occur elsewhere, we can construct a substitution $\theta = \{H \mapsto \lambda\overline{x_n}.\{x_i \mapsto t_i'\}t_0\} \cup S$, where $t_i' = \{\overline{X_o \mapsto u_o}\}t_i$, which is a solution to

$$\lambda x_0.u[H(t_1,\ldots,x_0,\ldots)]_p =^? \lambda x_0.t$$

Notice that θ is well-formed, as $\lambda\overline{y_m}.t_i$ does not contain (outside) bound variables. The other direction is simple, since x_0 does not occur elsewhere, i.e. not in an instance of $\lambda x_0.t$. □

Notice that the above procedure only helps deciding unification problems but does not imply that pre-unification or even unification is finitary.

Putting It All Together

Now we can combine the previous results. Recall that the remaining case is undecidable in general.

Theorem 5.2.8 *Assume s,t are λ-terms such that t is second-order, s is linear and s shares no variables with t. Furthermore, if $s|_p = F(\overline{t_n})$, then all $\overline{t_n}$ are either*

- *bound variables of arbitrary type or*
- *second-order ground terms of base type or*
- *second-order terms of base type without bound variables in $\mathcal{BV}(s,p)$.*

Then the unification problem $s =^? t$ is decidable.

Proof First apply Theorem 5.2.7 to the unification problem until s has no nested free variables. This argument can be applied repeatedly, as the lhs is linear and hence the substitutions of multiple applications do not overlap. Then Theorem 5.2.6 can be applied to decide this problem. □

A special case often considered (e.g. [Gol81]) is terms with second-order variables, but no bound variables. Then we get the following stronger result as an instance of Theorem 5.2.8:

Proposition 5.2.9 *Assume s,t are second-order λ-terms such that s is linear and shares no variables with t. Furthermore, s contains no bound variables. Then the unification problem $s =^? t$ is decidable.*

5.3 Relaxing the Linearity Restrictions

In this section we discuss unification problems with shared and repeated variables which were disallowed in the last section. The first result is an extension of higher-order patterns. The only known extension of higher-order patterns with unitary unification is due to Dale Miller [Mil91a]. Miller permits arguments to free variables that are patterns, but must have a bound variable as the outermost symbol. For instance, $\lambda x, y.P(x, y(f(x)))$ is permitted. The decidability result in the next section below allows second-order variables with patterns as arguments, as long as these variables occur only once.

The results in Section 5.3.2 show that unitary unification is easily lost when going beyond higher-order patterns. A further class with decidable unification is considered that does not subsume higher-order patterns but is interesting for some applications. For instance, unification of first-order terms with a term $\forall x.P(x) \Rightarrow P(x+1)$ is shown to be decidable.

5.3.1 Extending Patterns by Linear Second-Order Terms

We consider in the following an extension of higher-order patterns where subterms of the form $X(\overline{t_n})$ are permitted for some patterns $\overline{t_n}$ as long as X is second-order and does not occur elsewhere. This generalizes a result by Dowek [Dow93] which covers second-order terms with linear second-order variables, but without bound variables. Hence it does not subsume higher-order patterns.

We first need the following notation. A **linear second-order system** of equations is of the form

$$\overline{\lambda\overline{x_k}.X_n(\overline{t_{m_n}}) =^? \lambda\overline{x_k}.t'_n},$$

where all $\overline{X_n}$ are distinct and do not occur elsewhere and furthermore all $\overline{\lambda\overline{x_k}.t_{m_n}}$ and $\overline{\lambda\overline{x_k}.t'_n}$ are higher-order patterns. By abuse of notation, we write $\overline{t_{m_n}}$, avoiding nested bars.

For the next result recall from Section 4.1 that the elimination rule in System PT is not needed for completeness.

Theorem 5.3.1 *Unification of linear second-order systems is decidable.*

Proof We show that System PT for higher-order pre-unification terminates for linear second-order systems if the elimination rule is not used. For a system $S = \{\overline{\lambda\overline{x_k}.X_n(\overline{t_{m_n}}) =^? \lambda\overline{x_k}.t'_n}\}$ we use the following lexicographic termination ordering:

A: $|\mathcal{FV}(\overline{\lambda\overline{x_k}.t'_n}) \cup \mathcal{FV}(\overline{\lambda\overline{x_k}.t_{m_n}})|$

B: the multiset of sizes of $\overline{\lambda \overline{x_k}.t'_n}$

We show that the transformations of PT reduce this ordering. After a projection on the left (it may not occur on the right), an equation between two patterns is created. We consider solving this as an atomic operation (possibly reducing A). We maintain the invariant that the system remains a linear second-order system, which is easy to show. Hence we only have to consider the imitation and projection cases:

Imitation: in this case, the number of isolated variables on the left increases, but A remains constant and B is reduced after decomposition.

Projection reduces one equation to an equation between two patterns. Applying a solution of this equation (if it exists) to the remaining goals yields two cases as in Lemma 5.1.5: either A is reduced, or, if A remains the same, the substitution must not increase the size and thus B is reduced as one equation is removed.

□

Now we can show the desired result, where we represent the non-pattern terms in the unification problem by a substitution. This in fact yields a more general result, as the permitted non-pattern subterms may occur repeatedly. For instance, $f(X(a),X(a)) =^? p$ falls into this class, but $f(X(a),X(b)) =^? p$ does not, where p is a pattern.

Theorem 5.3.2 *Assume a substitution* $\theta = \{\overline{X_n \mapsto \lambda \overline{x_k}.X'_n(\overline{t_{m_n}})}\}$, *where* $\overline{t_{m_n}}$ *are patterns, and two patterns s and t. If all* $\overline{X'_n}$ *are distinct, second-order and do not occur elsewhere, then the unification problem* $\theta s =^? \theta t$*, is decidable.*

Proof It is sufficient to solve the pattern unification problem $s =^? t$ first, yielding a pattern substitution δ in a successful case. Then

$$\overline{\lambda \overline{x_k}.X'_n(\delta \overline{t_{m_n}}) =^? \lambda \overline{x_k}.\delta X_n}$$

is a second-order linear system and is decidable by Theorem 5.3.1. □

A typical application of Theorem 5.3.2 are contexts, which are often used to describe positions in terms. These are sometimes viewed as "terms with holes" and these holes are written as boxes. For instance, $f(\Box,a)$ and $C(\Box)$ can be viewed as contexts. It is clearly much more precise to express contexts by second-order terms. In particular, if a term has several different "holes". For instance, we would write $\lambda\Box.f(\Box,a)$ and $\lambda\Box.C(\Box)$ instead of the above and would let β-reduction perform the substitution for concrete values for "holes". Thus contexts can be modeled by linear second-order variables. With the above result, we have a method to determine if a first-order term t unifies with a (linear) context filled with some term. For instance, in order to find overlaps of two rules $l_i \to r_i$, $i = 0, 1$ in an abstract fashion, the equations $(\lambda\Box.C(\Box))l_i =^? l_{1-i}$, $i = 0, 1$ are to be solved. Notice that the last unification problems permit trivial solutions, e.g. $\{C \mapsto l_{i-1}\}$, which are not of interest here.

As another example, we can model term rewriting with contexts. Assume a rule $l \to r$. Checking if s is reducible by this rule is done by matching $(\lambda\Box.C(\Box))l$ with s. Similarly, for narrowing, as we see in the next chapter, unification of $(\lambda\Box.C(\Box))l$ with s is needed.

5.3.2 Repeated Second-Order Variables

We show in this section another decidability result for second-order unification that is tailored for a particular application. As we aim at relaxing the linearity conditions in results of the last section, we need several technical restrictions. Otherwise, this extension easily leads to infinitary unification problems. For instance, if ground terms are permitted as arguments to free variables, the following example shows that there exist infinitely many unifiers: recall the problem

$$F(f(a)) =^? f(F(a))$$

which has the solutions $\{F \mapsto \lambda x.f^n(a)\}$, $n \geq 0$.

Apart from the above example, equations of the form $F(t) =^? t'$, where F occurs in t', are unsolvable in most cases. We conjecture that the solvable cases are based on some symmetries. For instance, consider the equation

$$F(f(a,a)) =^? f(F(a), F(a)).$$

The solutions are of the form

$$\{F \mapsto \lambda x.x\}, \{F \mapsto \lambda x.f(x,x)\}, \{F \mapsto \lambda x.f(f(x,x), f(x,x))\}, \ldots$$

We conjecture that all solutions to equations $F(t) =^? s$ with $F \in \mathcal{FV}(s)$ are of such a form and can possibly be described by finite automata or grammars [Tho90].

Some interesting examples fall into this class. Consider unification with a typical induction scheme:

$$P(0), \forall x.P(x) \Rightarrow P(x+1) \vdash \forall x.P(x)$$

Typically in such formulas, some arguments to free variables are not bound variables, but ground (constructor) terms, here $P(x+1)$. Recall that the quantifier $\forall$ can be viewed as a second-order constant and that $\forall x.P(x)$ is nicer syntax for $\forall(\lambda x.P(x))$.

Unification of a term with a repeated free variable with some higher-order pattern permits infinitely many solutions: consider for instance

$$\forall x.f(P(x)) \Rightarrow P(f(x)) =^? \forall y.X(y) \rightarrow X(y)$$

which is similar to the above unification problem $F(f(a)) =^? f(F(a))$.

The main result of this section is that unification of such terms with (quasi) first-order terms is decidable. A term $\lambda\overline{x_k}.t$ is **quasi first-order** if t is first-order. For instance, $\lambda x.F(x), f(\lambda x.x)$ are not quasi first-order, but $\lambda x.f(x,P)$ is quasi first-order.[2] A simple property of quasi first-order terms we will use is the following: if t is quasi first-order, p is a pattern, and $\theta p = t$ then θ is quasi first-order on $\mathcal{FV}(p)$.

A **pattern with ground arguments** is a pattern with the exception that arguments to free variables are ground terms which contain at least one outside bound variable but no local binders. An example is $\lambda x.P(x+1,x)$, but $\lambda x.P(f(\lambda y.y))$ is not.

Lemma 5.3.3 *Assume $\lambda\overline{x_k}.t$ is a quasi first-order term, $\lambda\overline{x_k}.P(\overline{t_n})$ is a second-order pattern with ground arguments. Then the unification problem*

$$\lambda\overline{x_k}.P(\overline{t_n}) =^? \lambda\overline{x_k}.t$$

is decidable and, furthermore, if θ is a maximally general solution then θP is quasi first-order.

Proof Decidability follows from Theorem 5.3.1 as $\lambda\overline{x_k}.P(\overline{t_n}) =^? \lambda\overline{x_k}.t$ is a linear system and $P \notin \mathcal{FV}(\lambda\overline{x_k}.t)$ since t is first-order. We apply Imitation and Projection of System PT, except on elimination problems. This terminates by Theorem 5.3.1 for second-order linear systems. In case of a Projection, a matching problem of the form $\lambda\overline{x_k}.t_i =^? \lambda\overline{x_k}.t'$, where t' is quasi first-order, is created. This only has quasi first-order solutions, since t_i has no local binders.

Imitation may create elimination problems of the form

$$\lambda\overline{x_k}.P'(\overline{t_n}) =^? \lambda\overline{x_k}.X,$$

which can be solved by System EL. This yields the solution $\{P' \mapsto \lambda\overline{x_k}.X\}$, as all $\overline{t_n}$ contain bound variables. As P' cannot occur elsewhere in a linear system, the remaining unification problems do not change and the system

[2] This is more restrictive than the definition of quasi-first-order in [LS93, ALS94a].

remains linear. Thus decidability of the original unification follows and maximally general unifiers do exist. Furthermore, any solution for some X on the right is quasi first-order. Hence for any solution θ computed, $\lambda\overline{x_k}.\theta t$ is quasi first-order. This entails that θP must be quasi first-order as well, as $\theta P(\overline{t_n}) = \theta t$: if $\lambda\overline{x_k}.\theta P$ is not quasi first-order, then $\lambda\overline{x_k}.\theta P(\overline{t_n})$ cannot be quasi first-order, as all $\overline{t_n}$ are ground and of base type. □

Lemma 5.3.4 *Assume $\lambda\overline{x_k}.p$ is a higher-order pattern where no abstractions occur in p, and $\lambda\overline{x_k}.t$ is a quasi first-order term. Then maximally general solutions of the unification problem $\lambda\overline{x_k}.p =^? \lambda\overline{x_k}.t$ are quasi first-order.*

Proof We first construct a solution θ to the problem $\lambda\overline{x_k}.p =^? \lambda\overline{x_k}.t$. Then we show that θ cannot map some free variable in $\lambda\overline{x_k}.t$ to a term containing bound variables.

We apply the rules of system PU except on elimination problems. Since p has no locally bound variables and t is quasi first-order, we can assure the invariant that there are no bound variables except $\overline{x_k}$. Elimination problems are either of the form $\lambda\overline{x_k}.t_1 =^? \lambda\overline{x_k}.X$ or of the form $\lambda\overline{x_k}.P(\overline{y_m}) =^? \lambda\overline{x_k}.t_2$, and are solved by System EL. If an elimination problem is of the first form, it is clear that any solution for X must be first-order, as all bound variables in t_1 must be eliminated and since there are no locally bound variables on some lhs. For the second form of elimination problems, if solvable, the obvious solution $P \mapsto \lambda\overline{y_m}.t_2$ is quasi first-order. Thus any solution θ computed is quasi first-order.

This entails that $\theta\lambda\overline{x_k}.p = \theta\lambda\overline{x_k}.t$ is quasi first-order and hence θ must be quasi first-order for the free variables in p as well. □

Now we are ready for the main result of this section.

Theorem 5.3.5 *Assume $\lambda\overline{x_k}.p$ is a higher-order pattern where no abstractions occur in p, $\overline{\lambda\overline{x_k}.P(\overline{t_{m_n}})}$ are second-order patterns with ground arguments, $\overline{\lambda\overline{x_k}.P_n(\overline{y_{n_o}})}$ are patterns, and $\lambda\overline{x_k}.t$ is quasi first-order. Then the unification problem*

$$\lambda\overline{x_k}.p =^? \lambda\overline{x_k}.t, \overline{\lambda\overline{x_k}.P(\overline{t_{m_n}}) =^? \lambda\overline{x_k}.P_n(\overline{y_{n_o}})}$$

where $P \notin \mathcal{FV}(\lambda\overline{x_k}.p, \lambda\overline{x_k}.t)$ is decidable.

Proof The structure of the proof is as follows. We solve the equation $\lambda\overline{x_k}.p =^? \lambda\overline{x_k}.t$ as in Theorem 5.3.4, yielding a quasi first-order substitution θ. Then we show that the remaining equations can be solved under this substitution. We assume (w.l.o.g.) $P \notin \mathcal{FV}(\theta)$.

After solving $\lambda\overline{x_k}.p =^? \lambda\overline{x_k}.t$, there remain the following cases for the equations

$$\overline{\lambda\overline{x_k}.P(\overline{t_{m_n}}) =^? \lambda\overline{x_k}.\theta P_n(\overline{y_{n_o}})} \quad (5.1)$$

- All equations in (5.1) are flex-flex and thus have a solution. Otherwise
- some $P_i \in \mathcal{Dom}(\theta)$ exists. We solve the i-th equation $\lambda\overline{x_k}.P(\overline{t_{m_i}}) =^? \lambda\overline{x_k}.\theta P_i(\overline{y_{m_i}})$ as in Lemma 5.3.3, as $\lambda\overline{x_k}.\theta P_i(\overline{y_{m_i}})$ is quasi first-order. This yields a quasi first-order substitution θ' for P. Applying this to the remaining equations yields a set of equations with higher-order patterns only. Thus solving the remaining goals is decidable.

□

The last result applies directly to first-order theorem proving with additional induction schemes written as second-order formulas. For instance, consider a data structure for binary trees with the destructors *left_tree*, *right_tree*. Then a premise of an induction scheme for binary trees may read as

$$\forall x.P(left\_tree(x)) \wedge P(right\_tree(x)) \Rightarrow P(x)$$

Encoding a unification problem of a term $\forall x.t$ with the above scheme in the form required for the last result yields:

$$\begin{aligned} \forall x.P_1(x) \wedge P_2(x) \Rightarrow P_3(x) &=^? \forall x.t \\ \forall x.P(left\_tree(x)) &=^? \forall x.P_1(x) \\ \forall x.P(right\_tree(x)) &=^? \forall x.P_2(x) \\ \forall x.P(x) &=^? \forall x.P_3(x) \end{aligned}$$

Although we have found another decidable class of unification problems, the result also shows that it is increasingly complicated to describe these classes.

5.4 Applications and Open Problems

As mentioned in the introduction, higher-order unification is currently used in several theorem provers, programming languages, and logical frameworks. With the above results we can now develop simplified and somewhat restricted versions of the above applications that enjoy decidable unification. It should be mentioned that several systems such as Elf [Pfe91] and Isabelle[3] have already resorted to higher-order patterns, where unification behaves as in the first-order case.

The main restriction we use to achieve decidability is linearity. There is an interesting variety of applications where linearity is a common and sometimes also useful restriction. The main application and also the original motivation

[3] Isabelle still uses full higher-order pre-unification, if the terms are not patterns.

for this work is higher-order narrowing, which will be developed in the next chapter.

Recall for instance the rule

$$map(F, cons(X,Y)) \rightarrow cons(F(X), map(F,Y))$$

which has a linear pattern as the left-hand side. Interestingly, when coding functions such as *map* into predicates, as for instance done in higher-order logic programming [NM88], the head of the literal, e.g.

$$map_P(F, cons(X,Y), cons(F(X),L)) :- map_P(F,Y,L),$$

is not linear. However, when this rule is used only on goals of the form $map_P(t,t',Z)$, where Z is a fresh variable,[4] then the unification problem is decidable as it is equivalent to unification with a linear term. Thus our results also explain to some extent why unification in higher-order logic programming rarely diverges.

Another application area is type inference, which is mostly based on unification, whereby decidable static type inference for programming languages is desired. In many advanced type systems such as Girard's system F [GLT89] variables may range over functions from types to types, i.e. second-order type variables. In particular, Pfenning [Pfe88] relates type inference in the n-th-order polymorphic λ-calculus with n-th-order unification. As another example, for SML [MTH90] some restrictions avoid second-order unification problems in the module system. Thus progress in higher-order unification may help finding classes where type inference is decidable. However, non-unitary unification often means no principal (i.e. most general) type.

Other applications are described in the following.

Theorem Proving

Higher-order theorem provers often work with some form of a sequent calculus, where most rules have linear premises and conclusions e.g.

$$\frac{\Gamma \vdash A \quad \Gamma \vdash B}{\Gamma \vdash A \wedge B}$$

Furthermore, non-linear unification problems occur mostly with rewriting, e.g. with rules such as $P \wedge P \longrightarrow P$. For rewriting, however, only matching is required. Another interesting result for theorem proving was discussed in Section 5.3: unification of first-order terms with induction schemes of the form

$$\forall x.P(x) \Rightarrow P(x+1).$$

[4] Such variables are also called "output-variables" in [Red86].

Associative Unification

Unification modulo the law of associativity was an open problem for a long time, until Makanin [Mak77] showed its decidability.

It is known that associative unification can be embedded into higher-order unification, see e.g. [Pau94]. With Theorem 5.2.6, we can show the decidability of a class of problems that extends associative matching (which is rather trivial). The idea for the encoding is that function application is associative, for instance

$$\lambda x.(\lambda y.f(g(y)))h(x) = \lambda x.f(g(h(x))).$$

Thus associative lists are coded by functions, e.g. the list $[a,b]$ is coded by the term $\lambda x.cons(a,cons(b,x))$. It may seem that this representation for lists is clumsy, but it has the advantage that appending a list to another can be done by a single β-reduction. This representation has been discussed in [Hug86] and corresponds to the idea of difference lists in logic programming, as discussed in Section 8.1.

For instance, the matching problem

$$\lambda x.F(G(cons(c,x))) =^? \lambda x.cons(a,cons(b,cons(c,x)))$$

has the three solutions

$$\begin{array}{l}\{F \mapsto \lambda y.y, G \mapsto \lambda y.cons(a,cons(b,y))\},\\ \{F \mapsto \lambda y.cons(a,y), G \mapsto \lambda y.cons(b,y)\},\\ \{F \mapsto \lambda y.cons(a,cons(b,y)), G \mapsto \lambda y.y\}.\end{array}$$

As the left-hand side of the last matching problem is a second-order term, we can use Theorem 5.2.6 to decide associative unification problems where one side only has linear variables with ground arguments, e.g. the problem

$$\lambda x.F(G(cons(c,x))) =^? \lambda x.cons(Y,X(x))$$

is decidable. In this example, $\lambda x.F(G(cons(c,x)))$ represents all lists ending in c and $\lambda x.cons(Y,X(x))$ stands for a non-empty list.

5.4.1 Open Problems

We briefly mention some open problems for future examination. Is the unification of a pattern with a linear second-order term decidable? This might be equivalent to unification of two linear second-order terms. Another question is whether second-order flex-flex pairs can be solved finitely. This may be possible by extending EL. The counterexample in [Hue76] only gives a third-order pair with nullary unification.

The above unification problems are at least NP-hard, as they subsume second-order matching, which is NP-complete [Bax76]. Are they also NP-complete?

Is there any way to extend the results to the third-order case? Not an obvious one, since this would subsume third-order matching which may have infinitely many incomparable solutions. Another question is whether the particular strategy in Theorem 5.2.6 is really necessary for termination.

An interesting idea would be to combine the nice properties of higher-order patterns with the above decidability results. Assume we want to unify an arbitrary pattern with a second-order term. Then there are two overlapping decidable subclasses of this problem, i.e. pattern unification and Theorem 5.2.8. Apart from selecting the appropriate algorithm depending on the occasion, there is a more interesting way. The main problem for combining these is the linearity restriction in the above results. The idea is to linearize one of the terms to be unified. More precisely, if we unify an arbitrary pattern p with a second-order term t, we can first make p linear and add some equality constraints. Then we solve the unification problem and apply the solutions to the equality constraints.

For instance, if $p = \lambda y.f(X(y), X)$, we replace the unification problem $p =^? t$ by

$$\lambda y.f(X(y), Y) =^? t, X =^? Y.$$

Now if the first problem is solvable by θ and if t is a pattern as well, the resulting problem $\theta X =^? \theta Y$ is a pattern unification problem. With this construction, we can integrate both results. In fact, we can even decide further problems, although it seems difficult to describe this class. For instance, the unification problem

$$\lambda x.g(f(x, G_1), f(x, G_1)) =^? \lambda x.F(f(x, G)),$$

is similar to Example 4.1.3 but does not fall into any decidable class. We transform this into

$$\lambda x.g(f(x, G_1), f(x, G_2)) =^? \lambda x.F(f(x, G)), G_1 =^? G_2.$$

Solving first the linear version is equivalent to Example 4.1.3 and yields the solution $\{F \mapsto \lambda x.g(x,x), G_1 \mapsto G, G_2 \mapsto G\}$. It remains to solve the trivial equation $G =^? G$. Interestingly, repeating this linearization procedure may yield an algorithm for general second-order unification.

Chapter 6

Higher-Order Lazy Narrowing

This chapter discusses our main approach for solving higher-order equations: lazy narrowing. Lazy narrowing is a goal directed method for solving goals in a top-down or outside-in manner. It can be seen as a direct extension of higher-order unification by some narrowing rules. After starting with a general version of lazy narrowing, we develop refinements of lazy narrowing. Some of them also apply to equational reasoning, while the others are tailored towards functional-logic programming. In the former, we only assume terminating rewrite systems; in the latter we will in addition exploit properties of left-linear rewrite rules. Conditional equations are discussed in Section 6.4. Alternative approaches to narrowing are examined in the following chapter.

The results in Section 5.2 on the decidability of unification of two second-order λ-terms, where one term is linear, are one of the main motivations for this work: the unification problem needed for second-order narrowing is decidable if the left-hand sides of the rewrite rules are linear higher-order patterns. This result will be generalized to systems of equational goals in order to show that the unification rules do not diverge.

In the first, general version of lazy narrowing in Section 6.1, we show the principles of this approach and show completeness. In Section 6.2, we focus on narrowing with a terminating (general) higher-order rewrite system R. This permits focusing only on R-normalized solutions. As in the first-order case, this allows one to restrict the most unconstrained case of narrowing: narrowing at variable positions. Further refinements, which also require convergent rules, include simplification via rewriting into narrowing, as shown in Section 6.2.2. This is desirable, as simplification is a deterministic operation. A further important optimization is deterministic eager variable elimination, as examined in Section 6.2.3. In general, it is an open question if eager variable elimination is a complete strategy. In our setting, we can differentiate two cases of variable elimination, where elimination is deterministic in one

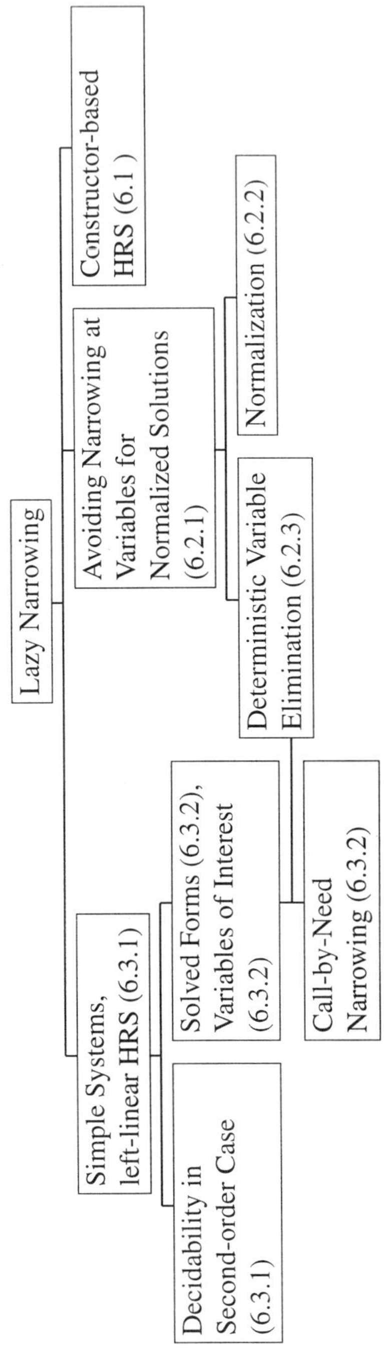

Figure 6.1: Dependencies of Lazy Narrowing Refinements

case.

In the following part, Section 6.3, we focus on refinements of lazy narrowing for functional-logic programming. In this setting the restriction to left-linear HRS is both common and useful. We show that for this setting a certain class of equational goals suffices. These are called Simple Systems (Section 6.3.1) and enjoy several nice properties. For instance, with the results on second-order unification of Section 5.2.1, we show that higher-order unification of second-order Simple Systems remains decidable, as it is in the first-order case. Furthermore, solved forms are much easier to detect than in the general case.

Combining the results for terminating systems with the properties of Simple Systems in Section 6.2.3 leads to an effective narrowing strategy, call-by-need narrowing. The basis for this strategy is a classification of the variables occurring in Simple Systems in Section 6.3.2. This allows one to recognize and to delay intermediate goals, which are only solved when needed.

As some of these refinements for lazy narrowing build upon others, we show these dependencies in Figure 6.1. Notice that all refinements can be combined in a straightforward way.

For both of the above settings, we show how conditional equations can be incorporated. Conditional rules are a common extension of term rewriting, useful in many applications. In Section 6.4.2 we argue that a class of conditional rules, called normal conditional rules, is sufficiently expressive in the higher-order case. In particular, we do not permit extra variables on the right sides of conditions. We show in Section 6.4.2 and Section 6.4.3 that the refinements developed for unconditional lazy narrowing can be extended to normal conditional narrowing. Finally, in Section 6.5, we discuss our narrowing approach and implications of the completeness results.

6.1 Lazy Narrowing

In this section, we introduce the central narrowing calculus. It will be refined later in order to reduce non-deterministic choices. Our setting for goal-directed lazy narrowing is as follows. For R-matching s with a ground term t, we start with a goal $s \rightarrow^? t$, where a substitution θ is a solution if $\theta s \xrightarrow{*}^R t$. This goal may be simplified to smaller goals by the narrowing rules, which include the rules of higher-order unification.

For the rules of System LN, shown in Figure 6.2, we need some notation. Let $s \overset{?}{\leftrightarrow} t$ stand for one of $s \rightarrow^? t$ and $t \rightarrow^? s$. For goals of the form $s \overset{?}{\leftrightarrow} t$, the rules are intended to preserve the orientation of $\overset{?}{\leftrightarrow}$. Recall that we extend the transformation rules on goals to sets of goals in the canonical way:

$\{s \to^? t\} \cup S \Rightarrow^\theta \overline{\{s_n \to^? t_n\}} \cup \theta S$ if $s \to^? t \Rightarrow^\theta \overline{\{s_n \to^? t_n\}}$.[1] For a sequence $\Rightarrow^{\theta_1} \ldots \Rightarrow^{\theta_n}$ of LN steps, we write $\stackrel{*}{\Rightarrow}^\theta$, where $\theta = \theta_n \ldots \theta_1$.

System LN for lazy higher-order narrowing essentially consists of the rules for higher-order unification plus the Lazy Narrowing rules. Observe that the first five rules in Figure 6.2 apply symmetrically as well, in contrast to the narrowing rule.

For a first impression of lazy narrowing, we start with a few examples. Assuming the rules

$$\begin{array}{lcl} map(F,[X|Y]) & \to & [F(X)|map(F,Y)] \\ map(F,[]) & \to & [] \\ \\ father(mary) & \to & john \\ father(john) & \to & art \end{array}$$

we solve the goal $R(mary) \to^? art$ by

$$\begin{array}{lll} R(mary) \to^? art & \Rightarrow_{LN} & \text{Narrowing at Variable,} \\ & & R \mapsto \lambda x.father(R_1(x)) \\ R_1(mary) \to^? john, art \to^? art & \Rightarrow_{LN} & \text{Deletion or Decomposition} \\ R_1(mary) \to^? john & \Rightarrow_{LN} & \text{Narrowing at Variable,} \\ & & R_1 \mapsto \lambda x.father(R_2(x)) \\ R_2(mary) \to^? mary, john \to^? john & \Rightarrow_{LN} & \text{Projection} \\ mary \to^? mary, john \to^? john & \stackrel{*}{\Rightarrow}_{LN} & \text{Solved by Deletion} \end{array}$$

Thus we get the solution $R \mapsto \lambda x.father(father(x))$. Notice that the trivial solution $R \mapsto \lambda x.art$ is also possible here, but it is easy to avoid by further constraints as e.g. in the following example.

Another, slightly more involved example is the following. We use functional evaluation in this example for brevity. Its completeness is shown later in Section 6.2.2.

$$\begin{array}{lll} map(F,[mary, john]) \to^? [john, art] & \stackrel{*}{\Rightarrow}_{LN} & \text{Evaluation,} \\ & & \text{Decomposition} \\ F(mary) \to^? john, F(john) \to^? art & \stackrel{*}{\Rightarrow}_{LN} & \text{Lazy Narrowing,} \\ & & F \mapsto \lambda x.father(H(x)) \\ H(mary) \to^? mary, john \to^? john, & & \\ \quad father(H(john)) \to^? art & \Rightarrow_{LN} & Projection, H \mapsto \lambda x.x \\ mary \to^? mary, john \to^? john, & & \\ \quad father(john) \to^? art & & \end{array}$$

[1]Some authors use multisets for this purpose, which is more precise but does not affect the results.

Deletion

$$t \to^? t \cup S \quad \Rightarrow \quad \{\}$$

Decomposition

$$\lambda\overline{x_k}.v(\overline{t_n}) \to^? \lambda\overline{x_k}.v(\overline{t'_n}) \quad \Rightarrow \quad \{\overline{\lambda\overline{x_k}.t_n \to^? \lambda\overline{x_k}.t'_n}\}$$

Elimination

$$F \overset{?}{\leftrightarrow} t \quad \Rightarrow^{\theta} \quad \{\} \text{ if } F \notin \mathcal{FV}(t) \text{ and } \theta = \{F \mapsto t\}$$

Imitation

$$\lambda\overline{x_k}.F(\overline{t_n}) \overset{?}{\leftrightarrow} \lambda\overline{x_k}.f(\overline{t'_m}) \quad \Rightarrow^{\theta} \quad \{\overline{\lambda\overline{x_k}.H_m(\overline{\theta t_n}) \overset{?}{\leftrightarrow} \lambda\overline{x_k}.\theta t'_m}\},$$
where $\theta = \{F \mapsto \lambda\overline{x_n}.f(\overline{H_m(\overline{x_n})})\}$
with fresh variables $\overline{H_m}$

Projection

$$\lambda\overline{x_k}.F(\overline{t_n}) \overset{?}{\leftrightarrow} \lambda\overline{x_k}.v(\overline{t'_m}) \quad \Rightarrow^{\theta} \quad \{\lambda\overline{x_k}.\theta t_i(\overline{H_p(\overline{t_n})}) \overset{?}{\leftrightarrow} \lambda\overline{x_k}.v(\overline{\theta t'_m})\},$$
where $\theta = \{F \mapsto \lambda\overline{x_n}.x_i(\overline{H_p(\overline{x_n})})\}$,
$\overline{H_p : \tau_p}$, and $x_i : \overline{\tau_p} \to \tau$
with fresh variables $\overline{H_p}$

Lazy Narrowing with Decomposition

$$\lambda\overline{x_k}.f(\overline{t_n}) \to^? \lambda\overline{x_k}.t \quad \Rightarrow \quad \{\overline{\lambda\overline{x_k}.t_n \to^? \lambda\overline{x_k}.l_n}, \lambda\overline{x_k}.r \to^? \lambda\overline{x_k}.t\},$$
where $f(\overline{l_n}) \to r$ is an $\overline{x_k}$-lifted rule

Lazy Narrowing at Variable

$$\lambda\overline{x_k}.H(\overline{t_n}) \to^? \lambda\overline{x_k}.t \quad \Rightarrow^{\theta} \quad \{\overline{\lambda\overline{x_k}.H_m(\overline{\theta t_n}) \to^? \lambda\overline{x_k}.l_m}, \lambda\overline{x_k}.r \to^? \lambda\overline{x_k}.\theta t\},$$
where $f(\overline{l_m}) \to r$ is an $\overline{x_k}$-lifted rule
and $\theta = \{H \mapsto \lambda\overline{x_n}.f(\overline{H_m(\overline{x_n})})\}$
with fresh variables $\overline{H_m}$

Figure 6.2: System LN for Lazy Narrowing

The last goals are easily solved by evaluation and deletion. Composing the partial bindings yields the solution $F \mapsto \lambda x.father(x)$. Observe how in the last examples Lazy Narrowing at Variable is used to compute solutions for functional variables. Although this rule is very powerful, it also has a high degree of non-determinism as it is quite unrestricted. We will show later how to restrict this rule.

To formalize the idea behind lazy narrowing and to approach the completeness result, we sketch a lazy narrowing computation more generally: for solving a goal $h(t_1,t_2) \to^? v(X,Y)$, we either simplify the goal to the goals $t_1 \to^? X$ and $t_1 \to^? Y$ if $h = v$, or apply a narrowing step at the root in a lazy fashion. That is, assuming a rule $h(a,Z) \to g(b)$, we transform the above goal to

$$\{t_1 \to^? a, t_2 \to^? Z, g(b) \to^? v(X,Y)\}.$$

For a solution $\theta h(t_1,t_2) \to^? \theta v(X,Y)$ not the first rewrite step is modeled by a narrowing step, but the first outermost one. Assume this is the rewrite step to t' in

$$\theta h(t_1,t_2) \xrightarrow{*} \theta h(a,Z) \longrightarrow_{\varepsilon}^{l \to r} t' \xrightarrow{*} \theta t.$$

Now the purpose of the goals $t_1 \to^? a, t_2 \to^? Z$ is easy to see: the rewrite steps in $\theta h(t_1,t_2) \xrightarrow{*} \theta h(a,Z)$ are modeled by these two goals.

In the last example, Z does not occur on the right-hand side of the rule $h(a,Z) \to g(b)$. Speaking in programming terminology, it is not necessary to "evaluate" the term t_2, here to Z. This corresponds to lazy evaluation, as t_2 can be reducible. The reason for this is that lazy narrowing, in its simple form, is also complete for reducible solutions, which makes it possible to model lazy evaluation.

It should be noted that the notion of laziness in Lazy Narrowing not only serves for lazy evaluation as in lazy or non-strict languages, but also to lazy instantiation of free variables. Intuitively, this means that instantiations are only performed when needed. This distinction will become clear later, e.g. in Section 8.1.

Before we present the completeness proof, we present a new rule which unifies the two narrowing rules of LN and simplifies the proofs. To model the two rules, it suffices to consider the step

$$s \to^? t \Rightarrow s \to^? \lambda \overline{x_k}.l, \lambda \overline{x_k}.r \to^? t$$

where $l \to r$ is an appropriately lifted rule. We also call this the **General Lazy Narrowing** rule. To obtain Lazy Narrowing with Decomposition, we can assume that s is of the form $\lambda \overline{x_k}.f(\ldots)$ and thus Decomposition applies after the General Lazy Narrowing rule. For Narrowing at Variable, s is of the form $\lambda \overline{x_k}.H(\ldots)$ where Imitation applies after the rule application. The difference between this generalized rule and the two special ones is that Decomposition/Imitation follow directly, without any other choice.

The completeness proof of system LN is built upon the completeness proof of higher-order unification in a modular way: the termination ordering is a lexicographic extension of the one in Theorem 4.1.7. As in pre-unification, we do not solve systems completely, but terminate if only flex-flex pairs remain.

Theorem 6.1.1 (Completeness of LN) *If $s \rightarrow^? t$ has solution θ, i.e. $\theta s \xrightarrow{*}^R \theta t$ for some GHRS R, then $\{s \rightarrow^? t\} \stackrel{*}{\Rightarrow}{}^{\delta}_{LN} F$ such that δ is more general modulo the newly added variables than θ and F is a set of flex-flex goals.*

Proof The proof proceeds by induction on the following lexicographic termination ordering on $(\overline{G_n}, \theta)$, where $\overline{G_n} = \overline{s_n \rightarrow^? t_n}$ is a system of goals with solution θ, i.e. $\overline{\theta s_n \xrightarrow{*} \theta t_n}$. Notice that a transformation not only changes $\overline{G_n}$, but also the associated solution has to be updated as in Theorem 4.1.7. The ordering assumes an arbitrary, but fixed reduction $\overline{\theta s_n \xrightarrow{*} \theta t_n}$.

- A: compare the multiset of length of the of R-reductions in each goal θG_i, if equal
- B: compare the multiset of sizes of the bindings in θ, if equal
- C: compare the multiset of sizes of the goals $\overline{G_n}$.

First consider the case that all goals are flex-flex pairs. Then the goals are considered solved. If not, we show that for any non flex-flex goal some rule applies that reduces the ordering.

Select some non flex-flex goal $s \rightarrow^? t$ from $\overline{G_n}$. In the base case for criteria A, that is $\theta s = \theta t$, some higher-order unification rule applies, as in Theorem 4.1.7. It is clear that this does not increase A and also approximates the solution θ. As the measures B and C are the ones used for Theorem 4.1.7, the above ordering is reduced.

Otherwise, there must be a rewrite step in $\theta s \xrightarrow{*} \theta t$. In the first case, assume there is no rewrite step at the root position in $\theta s \xrightarrow{*} \theta t$. Hence all terms in this sequence have the same root symbol. Then one of the unification rules must apply, as in the last case.

Now assume there are rewrite steps in $\theta s \xrightarrow{*} \theta t$ at root position. Then we consider the first of these, which we assume to be $\theta s \xrightarrow{*}^{\neq \varepsilon} \lambda \overline{x_k}.s' \longrightarrow^{l \rightarrow r}_{\varepsilon} \lambda \overline{x_k}.t_1$. Assume l is of the form $f(\overline{l_m})$. Hence θs is of the form $\lambda \overline{x_k}.f(\ldots)$. We have three cases. If s is of the form $\lambda \overline{x_k}.f(\overline{s_m})$, then Lazy Narrowing (with Decomposition) applies. If $s = \lambda \overline{x_k}.X(\overline{s'_p})$, then in case θX is of the form

$\lambda\overline{x_k}.x_i(\ldots)$, Projection applies as above. Otherwise, $\theta X = \lambda\overline{x_k}.f(\ldots)$ and Lazy Narrowing (at Variable) applies. We model both narrowing cases by the General Lazy Narrowing rule followed by immediate Decomposition/Imitation. In these cases $s' \to t_1$ must be an instance of $l \to r$, and there exists δ such that $\delta l \to \delta r = s' \to t'$. The General Lazy Narrowing rule yields the new goals

$$s \to^? \lambda\overline{x_k}.l, \lambda\overline{x_k}.r \to^? t$$

Since we model the first outermost reduction step in $\theta s \xrightarrow{*} \theta t$, it is safe to apply only Decomposition/Imitation on the first of the above goals (i.e. depending on the form of s, one of the two narrowing rules performs this task): no outermost rewriting can take place in $\theta s \to^? \lambda\overline{x_k}.\delta' l$. We can extend θ for the newly added variables: define $\theta' = \theta \cup \delta$. This is well defined, as we assume that $l \to r$ is renamed by an appropriate lifter. Thus θ' is a solution for the new goals, and coincides with θ on $\mathcal{FV}(s,t)$. The new goals have solutions with a smaller number of steps, thus reducing the termination ordering. □

A simple consequence of the last result is that System LN is complete for matching modulo convergent GHRS R. Unification can be encoded as shown in Section 6.5.1. Notice that we prove a more general result here, where solutions with $\theta s \xrightarrow{*} \theta t$ are considered. For equational matching, $\theta s \xrightarrow{*} t$ would suffice, but this is not strong enough to establish the result as above.

As discussed for System PT in Section 4.1, there are two sources of nondeterminism for such systems of transformations: which rules to apply and how to select the equations. As in Theorem 4.1.7, completeness does not depend on the goal selection, as each subgoal is independently solvable. Compared to pure higher-order unification, there is an important difference as the Elimination and Decomposition rules are, in general, not deterministic any more, as they may overlap with the narrowing rules.

It is interesting to compare LN with recent work on first-order lazy unification in [Han94c]. Restricting our system to the first-order case almost yields the system presented in [Han94c] (with the difference that we consider oriented goals). For instance, the transformations in [Han94c] yield so-called quasi-solved systems, which correspond to systems of (first-order) flex-flex pairs. Notice that the Imitation rule coincides for the first-order case with "partial instantiations" in [Han94c] and with the "root imitation" rule in [Sny91].

Narrowing Rules for Constructors

In practice, GHRS often have a number of symbols, called constructors, that only serve as data structures. For constructor symbols, we can extract a few simple rules for Lazy Narrowing. The main advantage is that their application is deterministic, i.e. no other rule applications must be considered. The rules in Figure 6.3 cover the cases where the root symbol of the left side of a goal is a constructor. Notice that the second and third rules are only possible with

Deterministic Constructor Decomposition

$$\lambda\overline{x_k}.c(\overline{t_n}) \to^? \lambda\overline{x_k}.c(\overline{t'_n}) \quad \Rightarrow \quad \{\overline{\lambda\overline{x_k}.t_n \to^? \lambda\overline{x_k}.t'_n}\}$$

if c is a constructor symbol

Deterministic Constructor Imitation

$$\lambda\overline{x_k}.c(\overline{t_n}) \to^? \lambda\overline{x_k}.F(\overline{y_m}) \quad \Rightarrow^\theta \quad \{\overline{\lambda\overline{x_k}.\theta t_n \to^? \lambda\overline{x_k}.H_n(\overline{y_m})}\},$$

where $\theta = \{F \mapsto \lambda\overline{y_m}.c(\overline{H_n(\overline{y_m})})\}$
and $\overline{H_n}$ are new variables

Constructor Clash

$$\lambda\overline{x_k}.c(\overline{t_n}) \to^? \lambda\overline{x_k}.v(\overline{t'_m}) \quad \Rightarrow \quad \mathit{fail}$$

if $c \neq v$, where v is not a free variable
and c is a constructor symbol

Constructor Occurs Check

$$\lambda\overline{x_k}.F(\overline{t_n}) \overset{?}{\leftrightarrow} \lambda\overline{x_k}.t \quad \Rightarrow \quad \mathit{fail}$$

if F occurs in t on a rigid path
not below a defined symbol

Figure 6.3: Deterministic Constructor Rules

oriented goals, where evaluation proceeds only from left to right. In contrast, the direction is immaterial for the first one. The determinism of the rules in Figure 6.3 follows immediately from the definition of a constructor: if $\lambda\overline{x_k}.c(\theta\overline{t_n}) \xrightarrow{*} t$, then t will have the constructor c as the root symbol.

6.2 Lazy Narrowing with Terminating Rules

We examine in this section refinements for lazy narrowing for terminating HRS. This does not entail that narrowing terminates, but allows one to restrict the solutions considered to R-normalized substitutions. As in the first-order case, this yields many important optimizations. For convergent HRS R it is desirable to focus on R-normalized solutions, as for any substitution there is an equivalent R-normalized one.

Some of the optimizations generalize well-known ideas of the first-order case, e.g. normalization in Section 6.2.2. The results on eager variable elimination in Section 6.2.3 are however new and hold only as we work with directed goals.

6.2.1 Avoiding Lazy Narrowing at Variables

We show in this section that narrowing at variables or variable headed terms like $X(\overline{x_n})$ is not needed for R-normalized substitutions with an GHRS R. For patterns reducibility of a term $\theta X(\overline{x_n})$ implies that θ is not R-normalized by Lemma 4.3.5, hence violating the assumption. We conjecture that in practice, as in higher-order logic programming [MP92a], most terms at run time are patterns and hence narrowing at variables is not needed very often.

This result generalizes the first-order case, as for first-order terms narrowing at variable position is not needed. This is the main idea of narrowing with R-normalized solutions. It is an important optimization, as narrowing at variable positions is highly unrestricted and thus may create large search spaces.

For this result the restriction to innermost reductions is necessary. The following results apply to convergent rewrite rules, since for any solution there exists an innermost reduction in this case.

Definition 6.2.1 System **LNN** is defined as a restriction of system LN where Lazy Narrowing at Variable is not applied to goals of the form $\lambda\overline{x_n}.X(\overline{y_m}) \to^? t$ where $\lambda\overline{x_n}.X(\overline{y_m})$ is a higher-order pattern.

Completeness follows similar to Theorem 6.1.1:

Theorem 6.2.2 *If $s \to^? t$ has solution θ, $\theta s \xrightarrow{*}{}^R \theta t$ is an innermost reduction, and θ is R-normalized for some terminating GHRS R, then $\{s \to^? t\} \overset{*}{\Rightarrow}{}^{\delta}_{LNN} F$ such that δ is more general, modulo the newly added variables, than θ and F is a set of flex-flex goals.*

Proof The proof proceeds as in Theorem 6.1.1; in addition we have to show the invariant that the solutions associated with all (new) variables are normalized substitutions.

Assume a goal with normalized solution θ. In case of a Projection or Imitation, the partial binding computed maps a variable X to a higher-order pattern of the form $\lambda\overline{x_n}.v(\overline{H_m(\overline{x_n})})$. The new solution constructed (as in the proof of Theorem 4.1.7) maps the newly introduced variables $\overline{H_m(\overline{x_n})}$ to subterms of θX, which are in R-normal form. Hence all $\overline{\theta H_m}$ must be in R-normal form. For the Elimination rule, no new variables are introduced, thus the solution remains R-normalized.

The critical case is when new variables are introduced in the narrowing rule. The narrowing rule is used if there are rewrite steps in $\theta s \stackrel{*}{\longrightarrow} \theta t$ at root position. We consider the first of these, which we assume to be $\theta s \stackrel{*}{\longrightarrow} \lambda\overline{x_k}.s_1 \longrightarrow_{\varepsilon}^{l\to r} \lambda\overline{x_k}.t_1$. Hence $s_1 \to t_1$ must be an instance of $l \to r$, say with substitution δ, i.e. $\delta l = s_1$. As δ may not be in R-normal form, we show that it is possible to use (any normal form) $\delta\!\downarrow_R$ instead. Assume $\delta l = \delta f(\overline{l_n}) = s_1$. As the above reduction is innermost, all true subterms of s_1 must be in R-normal form. (Note that s_1 is not a variable.) Hence $\overline{\delta l_n}$ are in R-normal form. From $\delta \stackrel{*}{\longrightarrow} \delta\!\downarrow_R$ (and $\overline{l_n \stackrel{*}{\longrightarrow} l_n}$) we conclude by Lemma 4.3.4 that $\overline{\delta l_n \stackrel{*}{\longrightarrow} (\delta\!\downarrow_R) l_n}$. As $\overline{\delta l_n}$ are in R-normal form, $\overline{\delta l_n = (\delta\!\downarrow_R) l_n}$ follows.

Since we assume that $l \to r$ is renamed with new variables, $\theta' = \theta \cup \delta\!\downarrow_R$ is well defined. Then, after applying the General Lazy Narrowing rule, θ' is a solution of the resulting goal system and is furthermore R-normalized.

Finally, with the invariant that θ' is R-normalized, it is clear from the completeness proof of LN that LNN is complete as there can be no rewrite step in the solution θ of a goal $\lambda\overline{x_n}.X(\overline{y_m}) \to^? t$ as θX is in R-normal form. □

A few comments are in order:

- The restriction to HRS in the last result permits a simpler proof which does not require termination of R, as shown in [Pre95c]. Termination is only needed for the existence of $\delta\!\downarrow_R$, as δ cannot be use for the following reason: In case the left-hand sides are non-patterns, then solutions to new variables may be reducible in case of a lazy narrowing step. Assume for instance a is reducible for some GHRS R. If an instance $\delta f(G(\lambda x.b))$ of a left-hand side $f(G(\lambda x.b))$ is R-normalized, then δ need not be R-normalized: consider e.g. $\delta = \{G \mapsto \lambda x.x(a)\}$. The above proof shows that it is sufficient to use some normal form of the substitution δ.

- The above result implies the following optimization: if a goal

$$\lambda\overline{x_n}.X(\overline{y_m}) \to^? t,$$

where $\lambda\overline{x_n}.X(\overline{y_m})$ is a pattern, is unsolvable by pure unification, then we can immediately fail this search path. Since this is an elimination problem as considered in Section 5.1, this is decidable in the second-order case.

To show the effect of this narrowing refinement in the higher-order case, consider the goal:

$$\lambda x, y.F(x) \rightarrow^? \lambda x, y.sin(x)$$

The narrowing rules of LNN and the Elimination rule do not apply. The obvious solution is obtained by Imitation on *sin* and Projection for x. Notice that for descendants of goals of the form $\lambda \overline{x_n}.X(\overline{y_m}) \rightarrow^? t$, Lazy Narrowing at Variable does not apply, since such goals are transformed only to goals of this form. For a slight modification of the example,

$$\lambda x, y.F(x, y) \rightarrow^? \lambda x, y.sin(x),$$

both rules, Elimination and Imitation, have to be attempted. This is remedied in Section 6.2.3.

6.2.2 Lazy Narrowing with Simplification

Simplification by normalization of goals is one of the earliest [Fay79] and one of the most important optimizations. Its motivation is to prefer deterministic reduction over search within narrowing. Notice that normalization coincides with deterministic evaluation in functional languages. For first-order systems, functional-logic programming with normalization has been shown to be a more efficient control regime than pure logic programming [Fri85, Han92]. For instance, if two rewrite rules overlap, simplification can be very effective. As an example consider the goal $diff(\lambda x.F, X) \rightarrow^? 0$, which is immediately solved via simplification. Otherwise, if other rules are attempted, there can be a large search space.

The main problem of normalization is that completeness of narrowing may be lost. For first-order (plain) narrowing, there exist several works dealing with completeness of normalization in combination with other strategies (for an overview see [Han94b]).

Consider the following example for simplification, which builds upon the rules in Section 2.5. Consider for instance the goal

$$\begin{array}{ll} \{\lambda x.diff(\lambda y.1 * ln(F(y)), x) \rightarrow^? \lambda x.cos(x)/sin(x)\} & \stackrel{*}{\Rightarrow} \quad \text{Evaluation} \\ \{\lambda x.diff(\lambda y.F(y), x)/F(x) \rightarrow^? \lambda x.cos(x)/sin(x)\} & \end{array}$$

This goal can be solved easily, as shown in the next section. Without simplification, all other rules for *diff* are tried, all but one failing after a few steps.

There are many similar benefits of simplification. Recall from Section 6.1 that deterministic operations are possible as soon as the left-hand side of a goal has been simplified to a term with a constructor at its root. For instance, with the rule $f(1) \to 1$, we can simplify a goal $f(1) \to^? g(Y)$ by

$$\{f(1) \to^? g(Y), \ldots\} \Rightarrow \{1 \to^? g(Y), \ldots\}$$

and deterministically detect a failure. (Note that g may not be a constructor.)

In the following, we show completeness of simplification for lazy narrowing under some restrictions on the GHRS employed. The result is similar to the corresponding result for the first-order case [Han94c]. The technical treatment here is more involved in many respects due to the higher-order case. Using oriented goals, however, simplifies the completeness proof.

For oriented goals, normalization is only complete for goals $s \to^? t$, where θt is in R-normal form for a solution θ. For instance, it suffices if t is a ground term in R-normal form. This is usually no restriction as discussed in Section 6.5.1 and corresponds to the intuitive understanding of directed goals.

Definition 6.2.3 A **simplification step** on a goal $s \to^? t$ is a rewrite step on s, written as $\{s \to^? t\} \Rightarrow_{NLN} \{s' \to^? t\}$ if $s \longrightarrow^R s'$. **Normalizing Lazy Narrowing (NLN)**, is defined as the rules of LN plus arbitrary simplification on goals.

Observe that simplifying the right-hand sides is not desirable. It often means redoing work and may produce solutions repeatedly.

Theorem 6.2.4 (Completeness of NLN) *Assume a confluent GHRS R that terminates with order $<^R$. If $s \to^? t$ has solution θ, i.e. $\theta s \stackrel{*}{\longrightarrow}^R \theta t$ where θt and θ are R-normalized, then $\{s \to^? t\} \stackrel{*}{\Rightarrow}{}^{\delta}_{NLN} F$ such that δ is more general modulo the newly added variables, than θ and F is a set of flex-flex goals.*

Proof Let $<^R_{sub} = <^R \cup <_{sub}$. Assume $\overline{G_n} = \overline{s_n \to^? t_n}$ is a system of goals with solution θ, i.e. $\overline{\theta s_n \stackrel{*}{\longrightarrow}^R \theta t_n}$.

The proof proceeds by induction on the following lexicographic termination order on $(\overline{G_n}, \theta)$:

- A: $<^R_{sub}$ extended to the multiset of $\{\overline{\theta s_n}\}$,
- B: multiset of sizes of the bindings in θ,
- C: multiset of sizes of the goals $\overline{\theta G_n}$,
- D: $<^R$ extended to the multiset of $\{\overline{s_n}\}$.

By Theorem 4.3.12, item A is terminating. For the proof we need the invariant that all $\theta\overline{t_n}$ are R-normalized terms. This allows one to assume innermost solutions for each subgoal.

In the following, we show that normalization reduces this ordering and, furthermore, that for a non flex-flex goal some rule applies which reduces the ordering. In addition, we show in each of these cases that the above invariants are preserved. First, we select some non flex-flex goal $s \rightarrow^? t$ from $\overline{G_n}$. If none exists, the case is trivial.

We first consider the case where a simplification step is applied to a goal, i.e. $s \rightarrow^? t$ is transformed to $s' \rightarrow t$. We obtain $\theta s \stackrel{*}{\longrightarrow} \theta s'$ from Lemma 4.3.4. As θt is in R-normal form, confluence of R yields $\theta s \stackrel{*}{\longrightarrow} \theta s' \stackrel{*}{\longrightarrow} \theta t$. Thus θ is a solution of $s' \rightarrow t$. For termination, we have two cases:

- If $\theta s = \theta s'$, measures A through C remain unchanged, whereas D decreases.
- If $\theta s \neq \theta s'$ measure A decreases.

Clearly, the invariants are preserved.

If no simplification is applied, we distinguish two cases: if $\theta s = \theta t$, then we proceed as in pure unification. Similar to Theorem 4.1.7, one of the rules of higher-order unification applies. In case of the Deletion rule, measure A decreases. Decomposition and Imitation reduce A. Projection only decreases B, whereas A is unchanged.

As in Theorem 6.2.2, normalization of the associated solution is preserved. Decomposition and Imitation yield new right-hand sides. These are subterms of θt and are thus R-normalized.

In the remaining case, there must be a rewrite step in $\theta s \stackrel{*}{\longrightarrow} \theta t$. In the first case, assume there is no rewrite step at the root position in $\theta s \stackrel{*}{\longrightarrow} \theta t$. Hence all terms in this sequence have the same root symbol. Then, similar to the last case, one of the unification rules must apply.

Now consider the case with rewrite steps in $\theta s \stackrel{*}{\longrightarrow} \theta t$ at root position. Assume the first of these to be $\theta s \stackrel{*}{\longrightarrow} \lambda\overline{y_k}.s_1 \longrightarrow_{\varepsilon}^{l\rightarrow r} \lambda\overline{y_k}.t_1$, with the rule $l \rightarrow r$. Notice that $s_1 \rightarrow t_1$ must be an instance of $l \rightarrow r$. Then we select a rule as in the proof of Theorem 6.1.1. If one of the Lazy Narrowing rules applies, yielding subgoals of the form:

$$\overline{\lambda\overline{y_k}.s_m \rightarrow^? \lambda\overline{y_k}.l_m}, \lambda\overline{y_k}.r \rightarrow^? t$$

As there exists δ such that $s_1 = \delta l$ and $t_1 = \delta r$, we can extend θ to the newly added variables. As δ may not be in R-normal form we use $\delta\!\downarrow_R$ Theorem 6.2.2. We define $\theta' = \theta \cup \delta\!\downarrow_R$. It follows that θ' is in R-normal form.

As $\theta' s_i$ are subterms of $\theta' s$, the claim $\theta' s_i <^R_{sub} \theta' s$ holds, and $\theta' \lambda \overline{x_k}.r <^R_{sub} \theta' s$ follows from $\theta' s \stackrel{*}{\longrightarrow} \theta' \lambda \overline{y_k}.r$. Thus θ' is a solution of $\overline{s_m \rightarrow^? l_m}$ and $\lambda \overline{y_k}.r \rightarrow^? t$, that coincides with θ on $\mathcal{FV}(\overline{G_n})$. □

The termination ordering in this proof is rather complex. For instance, the last item in the ordering is needed in the following example: assume a goal $\lambda x.c(F(x,t)) \rightarrow^? \lambda x.c(x)$ with solution $\theta = \{F \mapsto \lambda x, y.x\}$. Here, normalization of t does not change the term $\theta \lambda x.c(F(x,t))$ and thus does not contribute to the solution.

6.2.3 Deterministic Eager Variable Elimination

Eager variable elimination is a particular strategy of general E-unification systems. The idea is to apply the Elimination rule as a deterministic operation whenever possible. That is, when elimination applies to a goal, all other rule applications are not considered.

Eager variable elimination is of great practical value. Consider for instance the example from the last section:

$$\begin{array}{lll} \{\lambda x.diff(\lambda y.1 * ln(F(y)),x) \rightarrow^? \lambda x.cos(x)/sin(x)\} & \stackrel{*}{\Rightarrow} & \text{Evaluation} \\ \{\lambda x.diff(\lambda y.F(y),x)/F(x) \rightarrow^? \lambda x.cos(x)/sin(x)\} & \stackrel{*}{\Rightarrow} & \text{Decomp.} \\ \{\lambda x.diff(\lambda y.F(y),x) \rightarrow^? \lambda x.cos(x), & & \\ \lambda x.F(x) \rightarrow^? \lambda x.sin(x)\} & & \end{array}$$

At this point, we have to chose which goal to solve first. Since on the second Elimination applies, we prefer this and in addition consider no further rules on the second goal. It remains the goal

$$\{\lambda x.diff(\lambda y.sin(y),x) \rightarrow^? \lambda x.cos(x)\}$$

This last goal can be solved easily, by evaluation and deletion as shown in Section 8.1.2.

Unfortunately, it is an open problem of general (first-order) E-unification strategies if eager variable elimination is still complete [Sny91]. A result for first-order, orthogonal rules was given in [OMI95]. Their setting is quite different as they do not assume termination. Interestingly, in [Han94c] the elimination is purposely avoided in a programming language context as it may copy terms whose evaluation can be expensive.

In our case, with oriented goals, we obtain more precise results by differentiating the orientation of the goal to be eliminated. As we consider oriented equations, we can distinguish two cases of variable elimination. In one case elimination is deterministic, i.e. no other rules have to be considered. In other words, eager variable elimination is complete in this case.

Theorem 6.2.5 *Lazy Narrowing with System NLN combined with eager variable elimination on goals $X \to^? t$ with $X \notin \mathcal{FV}(t)$ is complete for convergent GHRS R.*

Proof We show that the elimination of X reduces the termination ordering in the proof of Theorem 6.2.2 (see Theorem 6.2.4 for the ordering): as θ is R-normalized, there can be no rewrite step in $\theta X \to^? \theta t$. Thus $\theta X = \theta t$ follows. Hence for all other goals $s \to^? s'$, $\theta s \to^? \theta s'$ remains unchanged for an elimination step at $X \to^? t$. Binding X to t removes one equation and reduces measure A. □

Although we only show completeness of eager variable elimination in the setting of NLN, it is also complete for LNN, with essentially the same proof.

In the general case, variable elimination may copy reducible terms with the result that the reductions have to be performed several times. Notice that this case of variable elimination does not affect the reductions in the solution considered, as only terms in normal form are copied: θt must be in normal form.

Unfortunately, it is unclear if Elimination on goals of the form $t \to^? X$ is deterministic. There are however a few important cases where the result can be shown:

Theorem 6.2.6 *System LNN with eager variable elimination on goals $t \to^? X$, where t is either*

- *ground and in R-normal form or*
- *a pattern without defined symbols.*

is complete for convergent GHRS R

Proof In both cases it is clear that θt is in R-normal form for an R-normalized solution θ. Then elimination of X reduces the termination ordering in the proof of Theorem 6.1.1 as in Theorem 6.2.5. □

As LNN is complete for normalized solutions, the last result yields a refinement for System LNN.

Notice that this refinement only holds with directed goals. In an undirected setting, reductions in both directions are possible. For instance, with the HRS $f(X) \to a$ the equation $P =^? f(P)$ can be solved with $\{P \to a\}$.

6.2.4 Avoiding Reducible Substitutions by Constraints

Although system LNN restricts narrowing at variable positions, system LNN can still compute reducible substitutions. For instance, assume a rule $f(a) \to$

b and the goal $H(a) \rightarrow^? b$. Then Narrowing at Variable followed by one imitation step with $\{H \mapsto \lambda x.f(H_1(x))\}$ creates the two goals

$$H_1(a) \rightarrow^? a, b \rightarrow^? b.$$

Performing Imitation on the first goal with $\{H_1 \mapsto \lambda x.a\}$ yields the solution $\{H \mapsto \lambda x.f(a)\}$, which is clearly not normalized.

Elimination

$$F \overset{?}{\leftrightarrow} t \quad \Rightarrow^{\theta} \quad \{Irr(\theta F)\}$$
if $F \notin \mathcal{FV}(t)$ and t is a pattern
where $\theta = \{F \mapsto t\}$

Imitation with Constraints

$$\lambda\overline{x_k}.F(\overline{t_n}) \overset{?}{\leftrightarrow} \lambda\overline{x_k}.f(\overline{t'_m}) \quad \Rightarrow^{\theta} \quad \{\overline{\lambda\overline{x_k}.H_m(\overline{\theta t_n}) \overset{?}{\leftrightarrow} \lambda\overline{x_k}.\theta t'_m)} \cup Irr(\theta F)\}$$
where $\theta = \{F \mapsto \lambda\overline{x_n}.f(\overline{H_m(\overline{x_n})})\}$
and $\overline{H_m}$ are new variables

(General) Lazy Narrowing with Constraints

$$\lambda\overline{x_k}.s \rightarrow^? \lambda\overline{x_k}.t \quad \Rightarrow \quad \{\lambda\overline{x_k}.s \rightarrow^? \lambda\overline{x_k}.l, \lambda\overline{x_k}.r \rightarrow^? \lambda\overline{x_k}.t, Irr(\mathcal{FV}(l))\}$$
where $l \rightarrow r$ is an $\overline{x_k}$-lifted rule

Constraint Failure

$$Irr(\overline{t_n}) \quad \Rightarrow \quad fail \quad \text{if some } t_i \text{ is } R\text{-reducible}$$

Figure 6.4: Rules for Lazy Narrowing with Constraints (LNC)

We can restrict the search for normalized substitutions further by adding constraints as shown in Figure 6.4. The idea of these constraints is to detect reducible substitutions. In the following, we show how to add such constraints in a safe way. Note that we use a version of the General Lazy Narrowing rule for simplicity. In this narrowing rule, we can avoid a trivial solution $\{H \mapsto \lambda\overline{x_n}.l\}$ to the goal $\lambda\overline{x_n}.H(\overline{t_n}) \rightarrow^? \lambda\overline{x_n}.l$, which leads to a reducible solution. Observe that in the first-order case this trivial solution is always possible, unlike in the higher-order case.

Similarly, we can add a constraint in the Imitation rule, if some variable X is partially instantiated by a term $f(\overline{X_n})$. Then for all computed substitutions θ the term $\theta f(\overline{X_n})$ must not be reducible. We denote these constraints by $Irr(\overline{t_n})$, with the intended meaning that $\overline{t_n}$ are not R-reducible.

The important invariant to preserve is that the terms in the constraints are patterns. In essence, the constraints only hold approximations for some variable of the solution to be computed. If the terms are non-patterns, reducibility of a term t in a constraint $Irr(t)$ does not imply that the solution considered, i.e. θt, is reducible. For patterns, however, Lemma 4.3.5 shows that θt is reducible, if t is.

Definition 6.2.7 We define System LNC by replacing the appropriate rules of System LN with the rules in Figure 6.4.

Recall that the Elimination rule is not necessary for completeness of both PT and LN. Thus it is possible to restrict the Elimination rule to patterns in LNC without losing completeness. System LNC can be viewed as a refinement of System LNN with some additional constraints. If some constraint is added, it is clear that the term must not be R-reducible. Furthermore, with the restriction of the elimination rule to patterns it follows, as for Theorem 4.1.5, that all computed substitutions are patterns. Thus completeness of LNC for normalized solutions follows easily. We will see later that this restriction for the Elimination rule is fulfilled for a strategy which we examine in Section 6.3.2.

It may seem that checking the reducibility constraints is costly, but with normalized substitutions many redundant narrowing attempts can be avoided early. This has been argued similarly for LSE narrowing [BKW93], where also many reducibility conditions have to be checked.

Notice that System LNC may introduce redundant constraints that lead to redundant checks, e.g. if $Irr(t)$ is added and t is a subterm of an existing constraint. The idea of this section is to show when it is possible to add irreducibility constraints. In applications it may be interesting to add constraints only selectively, as reducibility checks create overhead.

6.3 Lazy Narrowing with Left-Linear Rules

This section examines refinements for left-linear HRS. We will show that this class facilitates several optimizations not possible for general HRS. As left-linearity is common in functional(-logic) programming languages, this setting is of particular interest. For instance, the core of common functional languages such as SML or Haskell can be modeled by left-linear higher-order rewrite rules. For an overview of the development, we refer again to Figure 6.1. The starting point is a restricted class of goals that suffices for lazy narrowing with left-linear HRS. The refinements culminate in a call-by-need strategy.

6.3.1 An Invariant for Goal Systems: Simple Systems

In this section we introduce a particular class of goal systems, Simple Systems, with several interesting properties. For instance, the occurs check is not needed and it is easy to check if the system is solved. We show that this class is closed under the rules of LN for a left-linear HRS R. Furthermore, in the second-order case, the syntactic solvability (w.r.t. the conversions of λ-calculus) is decidable for systems of this class.

The invariant of Simple Systems allows for further optimizations, e.g. a closer analysis of the variables involved and eager variable elimination (Section 6.3.2). The properties are not specific to the higher-order case and apply to first-order systems as well. This holds particularly for results on solvability checks in the next section, which can be expensive in an actual implementation.

To introduce Simple Systems, we first define an ordering on goals:

Definition 6.3.1 We write $s \rightarrow^? s' \ll t \rightarrow^? t'$, if $\mathcal{FV}(s') \cap \mathcal{FV}(t) \neq \{\}$.

This ordering links goals by the variables occurring: e.g. $t \rightarrow^? f(X) \ll X \rightarrow^? s$. If $G_i \ll G_j$ we say there is a **connection** between these two goals. If two goals have no connection, then they are called **parallel**.

The following properties are essential for Simple Systems.

Definition 6.3.2 A system of goals $\overline{G_n} = \overline{s_n \rightarrow^? t_n}$ is called **cycle free** if the transitive closure of $\ll$ is a strict partial ordering on $\overline{G_n}$ and **right isolated** if every variable occurs at most once on the right-hand sides of $\overline{G_n}$.

Now we are ready to define Simple Systems:

Definition 6.3.3 A system of goals $\overline{G_n} = \overline{s_n \rightarrow^? t_n}$ is a **Simple System**, if

- all right-hand sides $\overline{t_n}$ are patterns,
- $\overline{G_n}$ is cycle free, and
- $\overline{G_n}$ is right isolated.

For instance, to solve a matching problem $s \rightarrow^? t$ we may assume (w.l.o.g.) that t is ground, thus the system is simple. A similar invariant for first-order horn clauses with equality was presented in [GLMP91]. For first-order lazy narrowing in [DMS92] right isolation is used, but the invariant there is not completely formalized. Another fragment of Simple Systems, i.e. the ordering on goals, is used in [CF91] to locate simplification steps after instantiation of variables.

The following properties of variables in Simple Systems follow easily from the definition:

Lemma 6.3.4 *Assume a Simple System $\overline{G_n}$ and a goal $G_i = (s \rightarrow^? t)$. Then*

- $\mathcal{FV}(s) \cap \mathcal{FV}(t) = \{\}$.
- *If $X \in \mathcal{FV}(s)$ and G_i is minimal w.r.t. $\ll^+$, then X occurs on no right-hand side.*
- *If $X \in \mathcal{FV}(t)$ and G_i is maximal w.r.t. $\ll^+$, then X occurs nowhere else.*

Solving a single goal $l \rightarrow^? r$ of a Simple System by pure unification is decidable in the second-order case by Theorem 5.2.1, since r is a linear pattern and l and r share no variables. We extend this to goal systems in Section 6.3.1. Notice that in a Simple System, no occurs check is needed, e.g. $P \rightarrow^? c(P)$ cannot occur. This extends to the full system of goals since no cycles are allowed. For instance, a "hidden" occurs check, as in $\{P \rightarrow^? c(Q), Q \rightarrow^? P\}$, is impossible.

Simple Systems and Lazy Narrowing

The next theorem shows that Simple Systems are closed under the rules of LN for a left-linear HRS. For the Decomposition rule and the two narrowing rules, the proof follows easily from the form of the goals in Simple Systems and from the restriction on the rules. The imitation and projection bindings introduce new variables, but do not create cycles. The Elimination rule requires a case distinction. For instance, when eliminating a goal of the form $t \rightarrow^? P$, the variable P does not occur in any other goal on the right-hand side. Notice that the restriction to patterns on the right-hand side fits nicely with the results in Section 6.1: if the left side of a goal has a constructor outside, a deterministic operation applies.

Theorem 6.3.5 *Assume a left-linear HRS R. If $\overline{G_n}$ is a Simple System, then applying LN with R preserves this property.*

Proof We have the following cases if a goal G_i is transformed:

- Deletion: trivial.

- Decomposition: assume a goal $G_i = f(\overline{t_n}) \rightarrow^? f(\overline{t'_n})$ is decomposed to $\overline{G'_n} = \overline{t_n \rightarrow^? t'_n}$. Then there can be no connection between some goals in $\overline{G'_n}$. Each of these new goals has at most the connections of G_i and no others. Hence the system remains simple.

- Elimination: we have two cases, depending on the form of G_i:

  - Assume $G_i = X \rightarrow^? t$ and let $\{\overline{X'_m}\} = \mathcal{FV}(t)$. There can be at most one goal G_j with X on the right-hand side. When substituting X in G_j, only new connections to G_j are created that are already in $\ll^+$. Substituting t for X on some left-hand side does not introduce new connections as the variables $\overline{X'_m}$ may not occur on the right-hand side of some other goal. As t is a pattern the right sides remain patterns and thus the system remains simple.
  - If $G_i = t \rightarrow^? X$, then X may occur only on the left-hand side of some goals. Hence right isolation and the pattern property follow trivially. To show the absence of cycles, assume some $G_j = C'(X) \rightarrow u$ with $G_i \ll G_j$. We show that no new connections are added. For all G_k with $\theta G_k \ll \theta G_j$, where $\theta = \{X \mapsto t\}$, we have $G_k \ll^+ G_i \ll G_j$ as X may not occur in G_k on the right. Hence $\ll^+$ remains unchanged, as this argument holds for all such goals G_j.

- Imitation: an imitation binding of the form $\{X \mapsto \lambda\overline{x_k}.f(\overline{X_n(\overline{x_k})})\}$ does not change the $\ll$-ordering and furthermore the right-hand sides remain linear patterns. The remainder of this case follows as in the Decomposition case.

- Projection: as in the Imitation case, with the only difference that Projection may eliminate variables, thus removing connections.

- Lazy Narrowing: for replacing a goal $G_i = s \rightarrow^? t$ by the goals $s \rightarrow^? \lambda\overline{x_k}.l$ and $\lambda\overline{x_k}.r \rightarrow^? t$ we assume that the variables in $\lambda\overline{x_k}.l$ are new. Thus the right-hand sides remain patterns with right isolated variables. Then $G_j \ll s \rightarrow^? \lambda\overline{x_k}.l$, iff $G_j \ll G_i$ and symmetrically for $\lambda\overline{x_k}.r \rightarrow^? t$. Thus no new connections are added and the system is simple.

□

With directed goals, left-linearity draws a line between equational matching, as done here, and unification: for equational unification, a non-linear rule

$X =^? X \to true$ must be added. For programming applications full unification is usually not needed. Furthermore, we will see that left-linear rules permit several optimizations.

For an implementation it is desirable that the goals are kept in an order compatible with $\ll$. Assume in an implementation that goals are kept in a list L which is in an order compatible with some ordering $<$. A transformation T on L **preserves the ordering** $<$, if applying T yields a list in an order compatible to $<$. The following property is easy to see:

Theorem 6.3.6 *System LN applied to a list of goals preserves the $\ll$-ordering for left-linear HRS.*

Proof by an analysis similar to the last proof. □

Solving Simple Systems

In the following, we show that solving second-order Simple Systems by unification is decidable. Furthermore, a particular solved form, which is equivalent to dag-solved form, is easy to detect in Simple Systems.

The following result implies that divergence in Simple Systems only stems from the lazy narrowing rules, as in the first-order case. This is important for practical applications. For instance, it is possible to determine if a Simple System has a syntactic solution before attempting a narrowing step.

For the next result we have to consider weakly second-order terms for the following reason: if a goal contains a second-order bound variable, lifting may yield a weakly second-order term.

Theorem 6.3.7 *Solving a weakly second-order Simple System $\overline{G_n}$ by unification is decidable and yields only a finite number of solutions.*

Proof We iteratively solve maximal (w.r.t. $\ll^+$) goals with LN. That is, if $s \to^? t$ is a maximal goal, then t is a linear pattern and the free variables in t may not occur elsewhere. Then solving this goal with PT (LN without the narrowing rules) terminates by Theorem 5.2.1 with a set of flex-flex pairs, all of which are of the form

$$\lambda \overline{x_k}.H(\overline{t_n}) \to^? \lambda \overline{x_k}.G(\overline{y_j}),$$

where G does not occur elsewhere. Such pairs can be finitely solved by Theorem 5.2.2. It remains to be seen that this solution preserves the property that the remaining system is simple: all solutions for $F \in \mathcal{FV}(\lambda \overline{x_k}.H(\overline{t_n}))$ are of the form $\{F \mapsto \lambda \overline{x_k}.F'(\overline{z_j})\}$, where $\{\overline{z_j}\} \subseteq \{\overline{x_k}\}$ and F' is a new variable of appropriate type. Hence, when applying this solution to the remaining equations, the system remains simple, as G does not occur elsewhere. □

Simple Systems have the advantage that it is easy to see if a system is in solved form, as we show next. In practice this means that checking whether the system is solved is less expensive. Furthermore, the occurs check is unnecessary as already pointed out.

Definition 6.3.8 A Simple System S is **simplified** if

$$S = \{X_1 \overset{?}{\leftrightarrow} t_1, \ldots, X_n \overset{?}{\leftrightarrow} t_n\}$$

and all $\overline{X_n}$ are distinct.

Theorem 6.3.9 *Simplified systems are solvable.*

Proof by induction on the number of goals. The base case is trivial. For the induction step, assume a maximal goal from a simplified system $\overline{G_n}$, say $G_n = X_n \overset{?}{\leftrightarrow} t_n$. Let $\theta = \{X \mapsto t_n\}$. We show that $\overline{\theta G_{n-1}}$ is simplified. There are two cases when applying the Elimination rule to G_n, depending on the form of G_n.

- If $G_n = t_n \rightarrow^? X_n$ then $\theta\overline{G'_{n-1}} = \overline{G'_{n-1}}$ as X is isolated and does not occur in $\overline{G'_{n-1}}$.

- In the other case, assume $G_n = X_n \rightarrow^? t_n$. Then

$$\overline{\theta G'_{n-1}} = \overline{X_{n-1} \overset{?}{\leftrightarrow} \theta t_{n-1}}$$

as all $\overline{X_n}$ are distinct. Thus the system $\theta G'_{n-1}$ is simplified.

Notice that the Elimination rule applies as $X_n \notin \mathcal{FV}(t_n)$. □

A simple corollary is the following.

Corollary 6.3.10 *A Simple System of the form* $\overline{\{t_n \rightarrow^? X_n\}}$ *is solvable.*

It is interesting to compare simplified systems to another well-known solved form: dag-solved form. This form is often used in the first-order case [JK91], but applies to our case as well.

Definition 6.3.11 A system of equations $\overline{X_n =^? t_n}$ is in **dag-solved form** if for all $i < j$, $X_i \neq X_j$ and $X_i \notin \mathcal{FV}(t_j)$.

A system of equations in dag-solved form can be described as

$$\begin{array}{lcl} X_1 & =^? & C_1(X_2,\ldots,X_n) \\ \vdots & & \\ X_i & =^? & C_i(X_{i+1},\ldots,X_n) \\ \vdots & & \\ X_n & =^? & C_n \end{array}$$

where all $\overline{X_n}$ are distinct and $\mathcal{FV}(\overline{C_n}) \cap \{\overline{X_n}\} = \{\}$.

Although simplified systems look very much like systems in dag-solved form, the $\ll$-ordering does not correspond to the ordering needed for dag-solved form. Let us show this by an example: the simplified system

$$Y \rightarrow^? f(X) \begin{array}{l} \ll X \rightarrow^? f(Z) \\ \ll g(X,Y) \rightarrow^? H \end{array}$$

is equivalent to the system (with unoriented equations)

$$H =^? g(X,Y), Y =^? f(X), X =^? f(Z)$$

This is the only ordering of the above goals which yields a system in dag-solved form. For this reason the following proof is tricky.

Theorem 6.3.12 *A system is simplified if and only if it is in dag-solved form (modulo orientation).*

Proof Clearly, orienting a system $\overline{X_n =^? t_n}$ in dag-solved form to $\overline{t_n \rightarrow^? X_n}$ yields a simplified system.

The other direction follows by induction on the number of goals. The base case is trivial. For the induction step, assume a maximal goal from a simplified system $\overline{G_n}$, say $G_n = X_n \overset{?}{\leftrightarrow} t_n$. Then by induction hypothesis $\overline{G_{n-1}}$ can be reordered (and reoriented) to dag-solved form, yielding $\overline{G'_{n-1}} = \overline{X'_{n-1} =^? t'_{n-1}}$. Then we again have two cases when applying Elimination to G_n, depending of the form of G_n.

- If $G_n = t_n \rightarrow^? X_n$ then X_n does not occur in $\overline{G'_{n-1}}$ as in Theorem 6.3.9. Thus $X_n =^? t_n, \overline{G'_{n-1}}$ is in dag-solved form.

- In the other case, $G_n = X_n \rightarrow^? t_n$. Then $\overline{G'_{n-1}}, X_n =^? t_n$ is in dag-solved form, as all variables in $\mathcal{FV}(t_n)$ are isolated and cannot occur in $\overline{G'_{n-1}}$.

□

6.3.2 A Strategy for Call-by-Need Narrowing

In this section we develop a new narrowing strategy for Simple Systems, assuming R-normalized solutions, which we call call-by-need narrowing.[2] In essence, we show that certain goals can safely be delayed, which means that computations are only performed when needed.

For this purpose, we first classify the variables occurring in Simple Systems in the next section. Then we show in Section 6.2.3 that the results on eager variable elimination from Section 6.2.3 can be extended in case of Simple Systems. This will reveal that in Simple Systems one case of variable elimination is not desirable, the other deterministic and always possible.

Variables of Interest

In the following, we classify variables in Simple Systems into variables of interest and intermediate variables. We consider initial goals of the form $s \to^? t$, and assume that only the values for the free variables in s are of interest, neither the variables in t nor intermediate variables computed by LN. For instance, assume the rule $f(a,X) \to g(b,X)$ and the goal $f(Y,a) \to^? g(b,a)$, which is transformed to

$$Y \to^? a, a \to^? X, g(b,X) \to^? g(b,a)$$

by Lazy Narrowing. Clearly, only the value of Y is of interest for solving the initial goal, but not the value of X.

This view is sufficient for solving matching problems, where the right side is ground. A simple example is encoding logic programs with predicates and with queries of the form $p(\ldots) \to^? \mathit{true}$. Alternatively, one may consider goals with free variables in the right-hand side that are considered as place holders for some value to be computed. For instance, if a function evaluates to pairs, we may only be interested in one component. Thus, for instance, the oriented query $s \to^? \mathit{pair}(0,X)$ may suffice.

The main result in this section is that Simple Systems allow us to identify variables of interest very easily. Furthermore, we will see how this distinction nicely integrates with our approach to eager variable elimination.

The interesting invariant we will show is that variables of interest only occur on the left, but never on the right-hand side of a goal. We first need to define the notion of variables of interest. Consider an execution of LN. We start with a goal $s \to^? t$ where initially the variables of interest are in s. This has to be updated for each LN step. If X is a variable of interest, and an LN

[2] This strategy was called needed lazy narrowing in [Pre95c].

step computes δ, then the free variables in δX are new variables of interest. With this idea in mind we define the following:

Definition 6.3.13 Assume a sequence of narrowing transformations $\{s \rightarrow^? t\} \stackrel{*}{\Rightarrow}^{\delta}_{LN} \{\overline{s_n \rightarrow^? t_n}\}$. A variable X is called a **variable of interest** if $X \in \mathcal{FV}(\delta s)$ and intermediate otherwise.

Now we can show the following result:

Theorem 6.3.14 *Assume a left-linear HRS R, a Simple System $\overline{G_n}$ such that $\overline{G_n} = \{\overline{s_n \rightarrow^? t_n}\}$ and a set of variables V with $V \cap \mathcal{FV}(\overline{t_n}) = \{\}$. If $\overline{G_n} \Rightarrow^{\delta}_{LN} \{\overline{s'_m \rightarrow^? t'_m}\}$, then $\mathcal{FV}(\delta V) \cap \mathcal{FV}(\overline{t'_m}) = \{\}$.*

Proof For all rules of LN, except the Elimination rule, the claim is trivial. For Elimination, consider first a goal of the form $t \rightarrow^? X$. In this case $X \notin V$ and X may not occur on any other right-hand side. Hence variables from V in t are only copied to some other left-hand side.

After the elimination of a goal $X \rightarrow^? t$ with $X \in V$, the right isolated free variables in t are in $\mathcal{R}ng(\delta)$, but do not occur in $\mathcal{FV}(\overline{t'_m})$. If $X \notin V$, nothing remains to show as $\mathcal{FV}(t) \cap V = \{\}$. □

Then the desired result follows easily:

Corollary 6.3.15 (Variables of Interest) *Assume a left-linear HRS R and assume solving a Simple System $s \rightarrow^? t$ with system LN. Then variables of interest only occur on the left, but never on the right-hand side of a goal.*

Notice that variables from the right may be shifted by the Elimination rule to some left-hand side.

The Two Cases of Variable Elimination

As we consider oriented equations, we can distinguish two cases of variable elimination and we will handle variable elimination appropriately in each case. In the first case,

$$X \rightarrow^? t,$$

the variable X can be a variable of interest. Thus the elimination of X is desirable for computational reasons and is deterministic for normalized solutions, as shown in Section 6.2.3. Notice that elimination is always possible on such goals in Simple Systems, as $X \notin \mathcal{FV}(t)$. In the context of Simple Systems we can refine the result for eager variable elimination in Section 6.2.3 by an additional failure case. Assume a goal $X \rightarrow^? t$ of a Simple System with an R-normalized solution θ. We have two cases:

- If t is in R-normal form, then elimination is deterministic by Theorem 6.2.5.
- If t is R-reducible, then the goal is unsolvable. As t is a pattern and the solution for X, i.e. θX, is R-normalized, $\theta X = \theta t$ must hold. This is impossible, as Theorem 4.3.5 entails that θt is reducible.

This observation shows the intuitive reason why Elimination is deterministic: in this case Elimination does not copy terms to be evaluated, t must be in normal form. In the other case of variable elimination, i.e.

$$t \rightarrow^? X,$$

elimination may not be deterministic and is not desirable, as we argue below.

Call-by-Need Narrowing

The results on intermediate variables and eager variable elimination in mind, we develop a new narrowing strategy. The idea is to delay goals of the form $t \rightarrow^? X$. This simple strategy has some interesting properties, which we will examine in the following.

We first view this idea in the context of a programming language. Let us for instance model the evaluation (or normalization) $f(t_1, t_2)\downarrow_R = t$ by Lazy Narrowing, assuming the rule $f(X, Y) \rightarrow g(X, X)$:

$$\{f(t_1, t_2) \rightarrow^? t\} \Rightarrow_{LN} \{t_1 \rightarrow^? X, t_2 \rightarrow^? Y, g(X, X) \rightarrow^? t\}$$

Now with the optimizations considered so far, variable elimination and normalization, we can model the following evaluation strategies.

Eager evaluation or call-by-value is obtained by performing normalization on the goals t_1 and t_2, followed by eager variable elimination on $t_1\downarrow_R \rightarrow^? X$ and $t_2\downarrow_R \rightarrow^? Y$. The disadvantage is that eager evaluation may perform unnecessary evaluation steps.

Call-by-name is obtained by immediate eager variable elimination on $t_1 \rightarrow^? X$ and on $t_2 \rightarrow^? Y$. It has the disadvantage that terms are copied, e.g. t_1 here as X occurs twice in $g(X, X)$. Thus expensive evaluation may have to be done repeatedly.

Call-by-need is an evaluation strategy that can be obtained by delaying the goals $t_1 \rightarrow^? X$ and $t_2 \rightarrow^? Y$, thus avoiding copying. Then t_1 and t_2 are only evaluated when X or Y are needed for further computation.

Call-by-need evaluation models equationally lazy evaluation with sharing copies of identical subterms [BvEG+87], i.e. the delayed equations may be viewed as shared subterms. It should be noted that the computed reduction may not be optimal as defined in [HL91], neither concerning the number of R-reductions nor of β-reductions. The notion of need considered here is similar to the notion of call-by-need in [Wad71, Lau93, AFM+95].

Let us now come back from evaluation to the context of narrowing. Consider for instance the Lazy Narrowing step with the above rule:

$$\{f(t_1,t_2) \to^? g(a,Z)\} \Rightarrow_{LN} \{t_1 \to^? X, t_2 \to^? Y, g(X,X) \to^? g(a,Z)\}$$

In contrast to evaluation as in functional languages, solving the goals $t_1 \to^? X, t_2 \to^? Y$ may yield many solutions. Whereas in functional languages, eager evaluation can be more efficient, this is unclear for solving equations or functional-logic programming. Thus we propose the following approach:

Definition 6.3.16 Call-by-need narrowing is defined as Lazy Narrowing where goals of the form $t \to^? X$ are delayed.

In the above example, decomposition on $g(X,X) \to^? g(a,Z)$ yields the goals $X \to^? a, X \to^? Z$. Then deterministic elimination on $X \to^? a$ instantiates X, thus the goal $t_1 \to^? a$ has to be solved, i.e. a value for t_1 is needed. In contrast, $t_2 \to^? Y$ is delayed.

This new notion of narrowing for Simple Systems and left-linear HRS is supported by the following arguments: **Call-by-need narrowing**

is complete, or safe, in the sense that when only goals of the form $\overline{t_n \to^? X_n}$ remain, they are solvable by Corollary 6.3.10. Since the strategy is to delay such goals, this result is essential. (This is not fully compatible with flex-flex pairs, as shown below.)

delays intermediate variables only. As shown in the last section, we can identify the variables to be delayed: a variable X in a goal $t \to^? X$ cannot be a variable of interest.

avoids copying, as shown above, variable elimination on intermediate variables possibly copies unevaluated terms and duplicates work. Thus intermediate goals of the form $t \to^? X$ are only considered if X is instantiated, i.e. if a value is needed.

Sharing, as modeled equationally in the Needed Lazy Narrowing strategy, is often considered on an implementational level only. In contrast, we have a more abstract view of sharing, which may lead to the same implementation: since each variable occurs only once on the right, it is sensible to view the delayed goals as a context of delayed terms. In an implementation, an intermediate variable can be associated with a pointer to the corresponding delayed goal. If the variable occurs repeatedly, this corresponds to sharing. This is explored in the next section.

We show this delaying of unnecessary computations by an example. Consider the goal

$$\lambda y.\mathit{diff}(\lambda x.\mathit{sin}(x) * F(x), y) \rightarrow^? \lambda y.0$$

w.r.t. the rules of Section 2.5. It can be simplified to

$$\lambda y.\mathit{diff}(\lambda x.\mathit{sin}(x)) * F(y) + \mathit{sin}(y) * \mathit{diff}(\lambda x.F(x), y) \rightarrow^? \lambda y.0$$

Now we apply narrowing with the rule $X + 0 \rightarrow X$:

$$\begin{aligned} \lambda y.\mathit{diff}(\lambda x.\mathit{sin}(x)) * F(y) &\rightarrow^? \lambda y.X(y), && (6.1) \\ \lambda y.\mathit{sin}(x) * \mathit{diff}(\lambda x.F(x), y) &\rightarrow^? \lambda y.0, && (6.2) \\ \lambda y.X(y) &\rightarrow^? \lambda y.0 && (6.3) \end{aligned}$$

This is followed by Elimination on X on the last goal and Narrowing at the first goal with $X' * 0 \rightarrow 0$:

$$\begin{aligned} &\lambda y.\mathit{diff}(\lambda x.\mathit{sin}(x)) \rightarrow^? \lambda y.X'(y), \\ &\lambda y.F(y) \rightarrow^? \lambda y.0, \lambda y.0 \rightarrow^? \lambda y.0, \\ &\lambda y.\mathit{sin}(x) * \mathit{diff}(\lambda x.F(x), y) \rightarrow^? \lambda y.0. \end{aligned}$$

Now the strategy for needed narrowing delays the first goal. As it is not needed for solving the other goals, it will never be handled. After solving the second goal by Elimination, which is deterministic here, the last two goals can be solved directly by Simplification and Decomposition. The final set of goals consists of the first goal of the above:

$$\lambda y.\mathit{diff}(\lambda x.\mathit{sin}(x)) \rightarrow^? \lambda y.X'(y) \qquad (6.4)$$

Its left-hand side is clearly not needed for the solution $F \mapsto \lambda x.0$ and is hence not evaluated, similar to lazy functional programming. Notice, however, that in some other solution, this term possibly has to be evaluated.

In practice, not only completeness but also (early) detection of failure is important. For instance, assume two goals

$$a \rightarrow^? X, b \rightarrow^? X,$$

where a and b are in normal form. Then by delaying both goals, the apparent unsolvability will never be detected. This will not occur with the above strategy in Simple Systems, as these are right isolated. Hence a variable X in a delayed goal $t \rightarrow^? X$ may occur on some left-hand side, but not on two right sides.

It is interesting to compare call-by-need narrowing with simplification, as shown in Section 6.2.2. First, simplification performs evaluation without any sharing and may both copy terms as well as evaluate unneeded terms. There are however cases, where evaluation is performed twice in two different search paths, which could be avoided by early simplification.

A Problem with Flex-Flex Pairs

Next we examine a problem that occurs in the higher-order case when call-by-need narrowing is employed and address solutions. In the above setting, two kinds of equations are delayed:

- Flex-flex goals of the form $\lambda \overline{x_k}.X(\overline{t_n}) \rightarrow^? \lambda \overline{x_k}.Y(\overline{y_m})$ and
- goals of the form $t \rightarrow^? X$.

A system of such goals can be unsolvable in general. Consider for instance the goal system

$$\begin{aligned} \lambda x.Y(x) &\rightarrow^? \lambda x.F, \\ \lambda x.f(x) &\rightarrow^? Y, \end{aligned}$$

which is unsolvable. Thus the delayed goals have to be solved by narrowing.

We will show two remedies. First, for working with rules as in functional languages, the problem cannot occur. Secondly, we can extend our notion of solved forms to account for this.

A solution for programming purposes is to assume fully-extended rules. A term where free variables are higher-order patterns and have all bound variables of the current scope as arguments, is called **fully extended**. A rewrite rule is **fully extended**, if the left-hand side is. For instance, this does not allow the rule

$$\mathit{diff}(\lambda y.F, X) \rightarrow 0.$$

Informally speaking, we do not allow complex λ-terms as data-structures in the left-hand sides. The rules still subsume functional languages programming, in which explicit binders do not occur in the left-hand sides.

The key idea is to maintain the additional invariant that the right-hand sides in goals are fully extended. Then, if a goal is flex-flex, it has a free variable on the right. Thus we need the following results:

Theorem 6.3.17 *Assume a left-linear, fully-extended HRS R. If $\overline{G_n}$ is a Simple System with fully extended right-hand sides, then applying LN with R preserves this property.*

Proof The proof proceeds as in Theorem 6.3.5. The two new ingredients are the following observations. First, for the narrowing rules, the new right hand side is a term lifted from some rewrite rule and is thus fully extended. For Imitation and Projection bindings are also fully extended terms, hence the property is stable. □

Corollary 6.3.18 *For a a Simple System with fully extended right-hand sides, flex-flex pairs are in solved form $t \rightarrow^? \lambda\overline{x_k}.X(\overline{x_k})$.*

From this solution for a practical, but limited case, we come back to the general setting. Even here, we conjecture that such cases are rare. Furthermore, we show that in many cases such goals are solvable:

Proposition 6.3.19 *Assume a second-order Simple System consisting of a set of flex-flex goals $\overline{G_m}$ and a set of goals $\overline{t_n \rightarrow^? X_n}$. Such a system is solvable if the following condition holds for all i and j: $t_i \rightarrow^? X_i \ll G_j$ implies that X_i is first-order.*

Proof The strategy of the proof is to eliminate goals from $\overline{t_n \rightarrow^? X_n}$ until only flex-flex goals remain. We show that each such Elimination transforms the goals into another set of goals of the above form. It clearly terminates, as the number of goals reduces.

For an elimination of $t_i \rightarrow^? X_i$, the variable X_i cannot occur on some other right-hand side. We consider two cases. If there is no G_j with $t_i \rightarrow^? X_i \ll G_j$, then X_i does not occur in some flex-flex goal and thus $\overline{G_m}$ does not change.

Otherwise, if

$$t_i \rightarrow^? X_i \ll G_j = \lambda\overline{x_k}.X(\overline{y_o}) \rightarrow^? \lambda\overline{x_k}.Y(\overline{z_m}),$$

then there are again two cases: if $X_i \neq X$, then G_j remains flex-flex, and the case is trivial. In the remaining case, we have $X_i = X$ and $n = 0$ as X and t_i are first-order by assumption. Binding X to t_i yields $\lambda\overline{x_k}.t_i \rightarrow^? \lambda\overline{x_k}.Y(\overline{z_m})$. As t is first-order, Lemma 5.1.4 for System EL applies, yielding the solution $\{Y \mapsto \lambda\overline{z_m}.t\}$. For simplicity, we only apply $\theta = \{Y \mapsto \lambda\overline{z_m}.Y'\}$ for a new variable Y', which yields an equation where Elimination applies. Thus θG_j is not flex-flex, but of the form of the $\overline{t_n \rightarrow^? X_n}$ goals. Furthermore, the other flex-flex goals remain flex-flex when applying θ. As this holds for all G_j, we obtain a smaller set of goals where the induction hypothesis applies. □

6.3.3 An Implementational Model

This section elaborates more operational and implementational aspects of call-by-need narrowing. In the following abstract model for call-by-need narrowing, goals are delayed in a context after Decomposition and Lazy Narrowing and are possibly reactivated by Elimination. The idea is to handle intermediate evaluations effectively and to detect deterministic operations on-the-fly.

The important step is to view the delayed goals for call-by-need narrowing as a "context" and to consider an intermediate variable as a pointer to a delayed term. This is possible for the following two reasons: intermediate variables can be characterized and, more importantly, variables can occur only once on the right and can hence be seen as a pointer to a (single) term. Thus we get contexts for free, i.e. we do not need any extra machinery. Notice that this works easier in the case of fully extended rules, where flex-flex pairs are in solved form. To formalize this, assume a set of delayed goals, or a context,

$$G_d = \overline{t_n \rightarrow^? X_n},$$

where $\overline{X_n}$ are guaranteed to be distinct, and a set of active goals

$$G_a = \overline{s_m \rightarrow^? s'_m}.$$

For an implementation, we assume that the intermediate variables $\{\overline{s_m}\} \cap \{\overline{X_n}\}$ have a "pointer" to their delayed goal in G_d. (Unfortunately the arrow in a delayed goal $t \rightarrow^? X$ gives the wrong direction for viewing this as a pointer.)

With this model in mind, we first examine the simpler narrowing rules on a goal from G_a. The idea of the following is to scan newly generated goals on-the-fly for deterministic operations.

Elimination on a goal $X \rightarrow^? t$ reactivates a delayed goal $t_i \rightarrow^? X$, if $X \in \{\overline{X_n}\}$.

Decomposition on a goal $\lambda\overline{x_n}.v(\overline{t_p}) \rightarrow^? \lambda\overline{x_n}.v(\overline{t'_p})$ creates new goals of the form $\overline{\lambda\overline{x_n}.t_p \rightarrow^? \lambda\overline{x_n}.t'_p}$.

Narrowing with Decomposition on a goal $\lambda\overline{x_n}.f(\overline{t_p}) \rightarrow^? s$ with a lifted rule $\lambda\overline{x_n}.f(\overline{l_p}) \rightarrow r$ creates new goals of the form $\overline{\lambda\overline{x_n}.t_p \rightarrow^? \lambda\overline{x_n}.l_p}, r \rightarrow^? s$.

For a set of new goals $\overline{G_n} = \overline{t_p \rightarrow^? t'_p}$, created by the Decomposition or Narrowing rule, we check if a deterministic operation is possible and if the goal is to be delayed. A goal from $\overline{G_n}$ can be of one of the following forms:

1. $\lambda\overline{x_l}.u(\overline{t_l}) \rightarrow^? \lambda\overline{x_l}.v(\overline{t'_l})$

2. $X \rightarrow^? \lambda\overline{x_l}.v(\overline{t'_k})$

3. $\lambda\overline{x_l}.u(\overline{t_k}) \rightarrow^? X$

When creating these goals, we check for deterministic simplification as follows. For the first form, we only check if a deterministic decomposition or if a constructor clash applies. Elimination is performed on goals of the second form. This may reactivate a delayed goal which is added to the new goals $\overline{G_n}$ and is checked as well. In the remaining case, goals of the third form are delayed. This (recursive) simplification procedure must terminate, as we only apply Elimination and Decomposition rules.

For the narrowing rules not considered above, we cannot hope for much preprocessing as above. Clearly, the Imitation, Projection and Narrowing at Variable rules are only needed on left-hand sides, which are no higher-order patterns. In this sense we have to pay the price when truly higher-order goals occur. Only Imitation and Narrowing at Variable on a goal create new goals with variable heads where in some cases Projection is the only operation that applies.

Let us show this activation of needed goals by the example above. In (6.1), the first equation is delayed, but when X is eliminated in the third goal, it is activated. Then, after a narrowing step, some part of the term is again delayed and remains until solved form is reached (6.4).

In some cases, it can however be advantageous to simplify the needed goal first. For instance, for a goal

$$\lambda y.\mathit{diff}(\lambda x.\mathit{sin}(x), y) \rightarrow^? X, \lambda y.\mathit{diff}(\lambda x.X(x), y) \rightarrow^? \lambda y. -1 * \mathit{sin}(y)$$

it is simpler to evaluate the first goal before guessing a rule for solving the second. Similarly, on delayed goals, deterministic operations, i.e. Simplification 6.2.2, Constructor Imitation (see Section 6.1) or Elimination (see Theorem 6.2.6), may apply and should be preferred. However, in general it is not clear what to attempt first.

6.4 Narrowing with Normal Conditional Rules

Adding conditions to equations is very common for rewrite systems. Although, at least in the first-order case, this may not increase the expressive power [BT87], conditions are often convenient. In the following sections, we develop conditional rewriting and narrowing with normal conditional rules.

Much research has been dedicated to narrowing with conditional equations. This has led to an abundance of different classes of conditional term rewrite systems and many different results. There exist various restrictions on the variables occurring in the conditions, for instance in [MH94] a hierarchy of four classes of conditional rules can be found. Combining these with the known strategies for (plain) narrowing led to an abundance of results in the first-order case, see for instance [MH94].

One of the problems with conditional rewriting is that termination of the associated rewrite relation does not imply the termination of conditional rewriting: rewriting the conditions proceeds recursively and may diverge without any actual reduction performed on the main goal. Thus most termination criteria for first-order conditional term rewriting need additional restrictions that assure that the reductions in the conditions are decreasing w.r.t. some termination ordering. For termination criteria that include conditional higher-order rewrite systems see for instance [LS93]. Another problem, addressed below, is that solving the conditions may require reducible substitutions, which renders many first-order strategies with plain narrowing incomplete.

We will point out in Section 6.4.2 that in our functional approach many of the above problems can be avoided due to the higher-order setting. The idea is that extra variables in the conditions are not needed for most programming constructs in the higher-order case.

6.4.1 Conditional Rewriting

In the following, we first introduce an unrestricted notion of conditional rules. For instance, the conditions may have variables not occurring in the rule itself, which are called extra variables. This is followed by a more restricted notion of rules, called normal conditional rules, which are suitable for programming purposes.

Definition 6.4.1 A **higher-order conditional rule** is of the form $l \to r \Leftarrow \overline{l_n \to r_n}$, where l is a higher-order pattern of base type and not η-equivalent to a free variable. A conditional HRS is a set of such rules and is abbreviated by **CHRS.**

There exist different notions of conditional rewriting. They differ in the way the equations are to be solved. Either they require $l_i \xrightarrow{*} r_i$, which is called normal equality [DO90], or logical equality w.r.t. R, which coincides with $l_i \downarrow r_i$ for confluent R and is often called join equality. The former is more tailored for programming languages, where evaluation is of interest, and usually the right-hand sides of the conditions are assumed to be ground R-normal forms, which we consider in the following section.

Definition 6.4.2 Assuming a rule $(l \rightarrow r \Leftarrow \overline{l_n \rightarrow r_n}) \in R$ and a position p in a term s in long $\beta\eta$-normal form, a **conditional rewrite step** from s to t is defined recursively as

$$s \longrightarrow_{p,\theta}^{l \rightarrow r \Leftarrow \overline{l_n \rightarrow r_n}} t \Leftrightarrow s \longrightarrow_{p,\theta}^{l \rightarrow r} t \wedge \overline{\theta l_n \stackrel{*}{\longrightarrow}^{R} \theta r_n}$$

Lifting rewrite rules over a set of variables extends to conditional rules by applying the lifter as well to the conditions.

Define the **length of a conditional reduction** as

$$\begin{aligned} len(s_1 \longrightarrow_{\sigma}^{l \rightarrow r \Leftarrow \overline{l_n \rightarrow r_n}} s_2 \stackrel{*}{\longrightarrow} s_n) &= 1 + len(s_2 \stackrel{*}{\longrightarrow} s_n) + \\ & \quad \Sigma_{i=1,\ldots,n} len(\sigma l_i \stackrel{*}{\longrightarrow} \sigma r_i) \end{aligned}$$

if $n > 1$ and $len(s_1) = 0$ if $n = 1$.

This notion of the length of a reduction reflects the problem with termination of conditional rewriting, since the rewrite steps for the conditions are included. As mentioned above, a conditional rewrite relation may itself terminate, but there may be reductions with infinite length.

It may seem tempting to examine conditional narrowing with normalized substitutions as in Section 6.2.1, but it is difficult to show that the solutions for the extra variables in the conditions are normalized. In the first-order case, this is a known problem as mentioned in the beginning of this section (see for instance [Han94b]). We therefore discuss conditional narrowing for a restricted class of rules in the following section, where the optimizations for unconditional rules of Section 6.1 can be adapted.

Definition 6.4.3 A **normal conditional HRS (NCHRS)** R is a set of conditional rewrite rules of the form $l \rightarrow r \Leftarrow \overline{l_n \rightarrow r_n}$, where $l \rightarrow r$ is a rewrite rule and $\overline{r_n}$ are ground R-normal forms.

Note that there is no difference between normal equality and joinability in our case as the right-hand sides of the rules are in ground R-normal form. Thus oriented goals suffice for proving the conditions since $\theta l_i \stackrel{*}{\longleftrightarrow} \theta r_i$ is equivalent with $\theta l_i \stackrel{*}{\longrightarrow} r_i$.

The definition of NCHRS may seem too restrictive, as no variables are allowed in the right sides of the conditions. As already discussed in Section 2.8.2, this is not needed for higher-order programming languages. We permit extra variables on the left sides of conditions, as these are needed to embed logic programs (for an example see Section 2.8.2). Extra variables on the right are often used to model local variables, which can be done here by "where" or "let" constructs of functional programming languages. These can easily be described by higher-order rules, such as

$$\text{let } X \text{ in } T \to T(X).$$

For instance, when writing a quicksort program, the main rule may be of the form

$$qs(O_{\leq}, S) \to append(qs(O_{\leq}, S_1), qs(O_{\leq}, S_2)) \Leftarrow split(O_{\leq}, S) \to (S_1, S_2)$$

where the S_i represent lists and (S_1, S_2) is a pair of lists. This is already not a first-order rule, as the ordering used for sorting is given as a parameter, here written as $O_{\leq}$. In our framework, we can write this as in a functional language as

$$\begin{aligned} qs(O_{\leq}, S) \to \; & \text{let } pair(s_1, s_2) \;=\; split(O_{\leq}, S) \\ & \text{in } \lambda s_1, s_2.append(qs(O_{\leq}, s_1), qs(O_{\leq}, s_2)), \end{aligned}$$

assuming a let rule for pairs, as shown in Section 8.1.

We believe that β-reduction is more appropriate than the instantiation of extra variables in the conditions. For instance, with depth-first search, as e.g. in Prolog, instantiations are recorded for possible backtracking. In contrast, β-reduction is a deterministic operation.

6.4.2 Conditional Lazy Narrowing with Terminating Rules

In this section, we discuss narrowing for a restricted class of conditional rewrite rules, which we argue to be sufficient for most purposes. For less restricted conditional rules we refer to [Pre95c]. We show in the following how to adapt the results for terminating rules to the conditional case.

Normalized Solutions and Variable Elimination

As we disallow variables on the right in conditions, it is easy to extend the results for narrowing with normalized solutions to normal conditional rules. For extra variables in the conditions it is often necessary to consider reducible solutions, making this important optimization impossible. In the following, we adapt the results of Sections 6.2 through 6.3 to conditional narrowing.

Definition 6.4.4 System **CLN** is defined as the unification rules of System LN plus the Conditional Lazy Narrowing rules in Figure 6.5.

For the following result, we disallow infinite conditional reductions. This is needed to construct innermost reductions for the proofs in the conditions

Conditional Narrowing with Decomposition

$$\lambda\overline{x_k}.f(\overline{t_n}) \to^? \lambda\overline{x_k}.t \quad\Rightarrow\quad \{\overline{\lambda\overline{x_k}.t_n \to^? \lambda\overline{x_k}.l_n}, \overline{\lambda\overline{x_k}.l'_p \to^? \lambda\overline{x_k}.r'_p}, \lambda\overline{x_k}.r \to^? \lambda\overline{x_k}.t\}$$

where $f(\overline{l_n}) \to r \Leftarrow \overline{l'_p \to r'_p}$ is an $\overline{x_k}$-lifted rule

Conditional Narrowing at Variable

$$\lambda\overline{x_k}.H(\overline{t_n}) \to^? \lambda\overline{x_k}.t \quad\Rightarrow^\theta\quad \{\overline{\lambda\overline{x_k}.H_m(\theta\overline{t_n}) \to^? \lambda\overline{x_k}.l_m}, \overline{\lambda\overline{x_k}.l'_p \to^? \lambda\overline{x_k}.r'_p}, \lambda\overline{x_k}.r \to^? \lambda\overline{x_k}.\theta t\}$$

if $\lambda\overline{x_k}.H(\overline{t_n})$ is not a pattern, $f(\overline{l_m}) \to r \Leftarrow \overline{l'_p \to r'_p}$ is an $\overline{x_k}$-lifted rule, and $\theta = \{H \mapsto \lambda\overline{x_n}.f(\overline{H_m(\overline{x_n})})\}$ with fresh variables $\overline{H_m}$

Figure 6.5: Rules for Conditional Lazy Narrowing (CLN)

as well. If one is only interested in such solutions, the requirement can be discharged. Later, this will be achieved by the slightly stronger restriction to decreasing rewrite rules. For a result without this restriction (in the lines of LN), we refer to [Pre95c]. Note that we view the conditional narrowing rules as composed of a Conditional Narrowing rule plus Decomposition/Imitation, as for the unconditional case.

Theorem 6.4.5 *Assume a convergent NCHRS R without infinite conditional reductions. If $s \to^? t$ has solution θ, i.e. $\theta s \xrightarrow{*}^R \theta t$, with an innermost reduction, and θ is R-normalized, then $\{s \to^? t\} \stackrel{*}{\Rightarrow}^{\delta}_{CLN} F$ such that δ is more general modulo the newly added variables than θ and F is a set of flex-flex goals.*

Proof proceeds as in Theorem 6.2.2, with a few modifications. We have to show the invariant that (intermediate) solutions are R-normalized. The problem here are new variables in the conditions of the rewrite rules. Thus, in the first part of the proof, we construct a new conditional reduction $\theta s \xrightarrow{*}^R \theta t$.

For a conditional rewrite step in $\theta s \xrightarrow{*}^R \theta t$ it is possible that the substitution is not R-normalized for some new variables in the reduction in a condition. Consider a rewrite step $\theta s \xrightarrow{*} s_1 \longrightarrow_{\delta}^{l \to r \Leftarrow \overline{l_n \to r_n}} s_2$ with $\delta l_n \xrightarrow{*} r_n$.

Let $V = \mathcal{FV}(\overline{l_n}) - \mathcal{FV}(l)$. As $\delta|_V$ may not be R-normalized, we construct an equivalent reduction $s_1 \longrightarrow_{\delta'}^{l \to r \Leftarrow \overline{l_n \to r_n}} s_2$, where $\delta' = \delta\downarrow_R$. Furthermore, we can assume that the reductions $\delta' l_i \stackrel{*}{\longrightarrow} r_i$ are innermost since R is convergent and all $\overline{r_n}$ are in ground R-normal form. This can be repeated recursively for all conditional reductions in $\theta s \stackrel{*}{\longrightarrow}^R \theta t$. The process terminates as there are no infinite conditional reductions.

Normalization for variables $X \in \mathcal{FV}(l)$ is shown as follows: As l is a pattern, all terms in $Im(\delta)$ are subterms of s_1 (modulo renaming, see Theorem 4.3.5). As the above reduction is innermost, all true subterms of s_1 must be in R-normal form. Hence δX must be R-normalized as well.

With this newly constructed rewrite sequence, we can show completeness similar to the proof of System LN in Theorem 6.2.2. We use induction on the following termination ordering for a system of goals $\overline{G_n} = \overline{s_n \to t_n}$ with solution θ.

- A: The sum of the lengths of the conditional R-reductions in all goals $\overline{\theta G_n}$.
- B: Multiset of the sizes of the bindings in θ.
- C: Multiset of the sizes of the goals $\overline{G_n}$.

Only the case with a lazy narrowing step differs from System LN: consider the first rewrite step at root position, which is assumed to be

$$\theta s \stackrel{*}{\longrightarrow} \lambda \overline{x_k}.s_1 \longrightarrow_{\varepsilon}^{l \to r \Leftarrow \overline{l'_o \to r'_o}} \lambda \overline{x_k}.t_1.$$

Hence $s_1 \to t_1$ must be an instance of $l \to r \Leftarrow \overline{l'_o \to r'_o}$ such that the conditions are solvable. Therefore, there exists a substitution δ with $\delta l = s_1$ and $\delta r = t_1$ such that $\overline{\delta l'_o \longrightarrow r'_o}$. Let m be the number of (conditional) reductions in $\overline{\delta l'_o \longrightarrow r'_o}$. Thus the size of this conditional rewrite step is $m+1$. Hence applying Conditional Narrowing reduces A, as one conditional reduction of size $m+1$ is replaced by new goals with conditional reductions of a total size m. As shown above, $\theta' = \theta \cup \delta$ is a R-normalized solution to the newly added goals. □

Note that we only show the last result for conditional HRS. The proof for unconditional GHRS in Theorem 6.1.1 cannot be adapted, since Lemma 4.3.4 is not available here. As no variables in the right-hand sides of the conditions are allowed, it is easy to see that the results for variables of interest, variable elimination of Section 6.3.2, and narrowing strategies of Section 6.3.2 hold in this context. Only simplification is more involved, as shown next.

Simplification for Conditional Narrowing

We show how the results for narrowing with simplification in Section 6.2.2 can be adapted to normal conditional narrowing. The basis for this is the restriction to normalized solutions, as elaborated above.

Definition 6.4.6 A termination ordering $<^R$ is a **decreasing termination ordering** for an NCHRS R, if $\theta l'_i <^R \theta l$ for any θ and for all $l_i \in \{\overline{l_n}\}$ and $l \to r \Leftarrow \overline{l_n \to r_n} \in R$.

Decreasing termination orders as defined here originate from the (first-order) definition in [DOS88] and imply termination of conditional rewriting (similar to [DOS88]). This is easy to show since for any rewrite step, the left-hand sides of the conditions are smaller in the ordering. Thus it suffices to consider a multiset of reductions. Then a conditional rewrite step performs one reduction and adds only smaller elements, i.e. the conditions, to the multiset.

Definition 6.4.7 System **NCLN** is defined as System CLN plus arbitrary simplification steps.

Theorem 6.4.8 *Assume a confluent NCHRS R with a decreasing termination ordering $<^R$. If $s \to^? t$ has solution θ, i.e. $\theta s \xrightarrow{*}^R \theta t$ where θt is in R-normal form, then $\{s \to^? t\} \overset{*}{\Rightarrow}{}^{\delta}_{NCLN} F$ such that δ is more general modulo the newly added variables than θ and F is a set of flex-flex goals.*

Proof is similar to Theorem 6.2.4, with a few changes and extra cases. As R is convergent, we can assume (w.l.o.g.) that the reduction $\theta s \xrightarrow{*}^R \theta t$ is innermost. Then we construct an innermost reduction in the proofs of the conditions as in Theorem 6.4.5, where solutions to all new variables are in R-normal form. We show completeness with the termination ordering and invariant used for NLN in Theorem 6.2.4. It suffices to consider a Conditional Lazy Narrowing step, all other cases are as in Theorem 6.2.4. Let $\overline{G_n} = \overline{s_n \to^? t_n}$ be a system of goals with solution θ. We merge the two narrowing rules (as in the General Lazy Narrowing rule) and consider just the following step, which is either followed by decomposition or imitation (for narrowing at variable):

$$
\begin{array}{rcll}
\{G_i\} & = & \{\lambda\overline{x_k}.s \to^? \lambda\overline{x_k}.t\} & \Rightarrow \\
\{\overline{G'_m}\} & = & \{\lambda\overline{x_k}.s \to^? \lambda\overline{x_k}.l, \overline{\lambda\overline{x_k}.l_o \to \lambda\overline{x_k}.r_o}, \lambda\overline{x_k}.r \to^? \lambda\overline{x_k}.t\} &
\end{array}
$$

Define $\theta' = \theta \cup \delta$, where δ is defined as in Theorem 6.4.5. Normalization of θ' follows as in Theorem 6.4.5. To complete the case, we only have to show two additional claims in order to extend the completeness result for NLN in Theorem 6.2.4 to NCLN:

- First, it is to assure that all $\lambda\overline{x_k}.\theta' r_o$ are R-normalized terms. This holds for the newly added conditions $\overline{\lambda\overline{x_k}.l_o \rightarrow^? \lambda\overline{x_k}.r_o}$, as $\lambda\overline{x_k}.\overline{r_o}$ are ground terms in R-normal form.

- The above narrowing step reduces measure A of Theorem 6.2.4, as $\overline{\theta' l_o <^R \theta s}$, for a decreasing ordering.

□

6.4.3 Conditional Lazy Narrowing with Left-Linear Rules

To carry over the results needed for the call-by-need narrowing strategy, all we show is that Simple Systems are invariant under the rules of CLN for a left-linear NCHRS R.

Theorem 6.4.9 *Assume a left-linear NCHRS R. If G is a Simple System then applying CLN with R preserves this property.*

Proof Building upon Theorem 6.3.5, we only consider the case of Conditional Lazy Narrowing. In this case, the right-hand sides of the conditions are ground terms and the new variables in the left sides of the conditions occur only on the left. Thus the system remains simple. □

Similarly, we get the following result as in Section 6.3.1:

Theorem 6.4.10 *System CLN with a left-linear NCHRS applied to goals in a list preserves the $\ll$-ordering.*

The only remaining issue is the problem with flex-flex pairs from Section 6.3.2. Since the right hand sides of the conditions are ground, they are trivially fully extended. Hence the solution for this case carries over and the implementational model is easy to extend for conditional rules.

6.5 Scope and Completeness of Narrowing

Before we conclude with higher-order lazy narrowing, we take a closer look at the kind of completeness results we have achieved. The completeness results are stated here in proof-theoretic terms with a clear operational model in mind. It is useful to compare and to discuss the extent of the corresponding completeness results.

In our approach, we only show completeness of narrowing w.r.t. solutions, sometimes only normalized substitutions. That is, for a goal $s \rightarrow^? t$, where t is ground, we consider solutions θ with $\theta s \xrightarrow{*} t$. (For most proofs this is generalized to $\theta s \xrightarrow{*} \theta t$.) We view this as the most general and basic concept of narrowing, as most of the common notions of completeness are easy to derive. For instance, for a convergent HRS R, this yields a complete algorithm for matching modulo the equational theory of R (for unification see below). In convergent theories, for any solution there exists an equivalent normalized one, thus our results suffice for complete R-unification or R-matching. For some results we explicitly require a convergent HRS and also give results tailored towards convergent HRS.

An alternative notion of completeness has been developed for programming language applications. Taking denotational semantics as the basis, two terms are equal, roughly speaking, if they can be evaluated to the same constructor term. This is called strict or continuous equality. For this notion of completeness w.r.t. denotational semantics [Red85], it suffices to consider only constructor based solutions, which are clearly normalized. This approach has led to implementations of functional-logic programming, see for instance [GLMP91, MNRA92]. It has also been extended to non-confluent HRS, which are used for non-deterministic programming [Huß93].

Strict equality can also be used with non-terminating rewrite rules, which has been claimed as an advantage of this approach. In contrast, we argue in Section 8.1.5 that non-terminating rules, as used in lazy functional languages, are not needed in logic programming.

Strict equality can be encoded with left-linear rules in our setting. We simply define a function $=_s$ that forces the evaluation to a constructor term. For instance, for natural numbers the rules

$$\begin{aligned} s(X) =_s s(Y) &\rightarrow X =_s Y \\ 0 =_s 0 &\rightarrow \mathit{true} \\ s(X) =_s 0 &\rightarrow \mathit{false} \\ 0 =_s s(X) &\rightarrow \mathit{false} \end{aligned}$$

suffice, assuming the constructors s and 0. This encoding works in a straightforward manner for first-order data types. It is however unclear how to extend strict equality to the higher-order case. There is a problem of higher-order strict equality with bound variables as e.g. in the goal $\lambda x.f(x) =_s \lambda x.f(x)$. One possible remedy is to assume an infinite set of rules $\lambda \overline{x_n}.x_i =_s \lambda \overline{x_n}.x_i$ for all n and i.

Interestingly, strict equality can be used to force functional evaluation. For instance, to evaluate a term t, a goal $t \rightarrow^? X$ may not be sufficient, as it is delayed in the call-by-need strategy. Evaluation is forced by the goal $t =_s X \rightarrow^? \mathit{true}$. Notice that only first-order objects are evaluated by this method, as in functional languages.

In the first-order case, our notion of plain narrowing (with some additional control strategy) is also called lazy narrowing (see e.g. [Han94a, LLR93]). Furthermore, lazy narrowing as defined here is called lazy unification in the first-order case [Han94c, MRM89]. Our naming conventions are based on some earlier works [Sny90, Höl88, Höl89].

6.5.1 Oriented versus Unoriented Goals

We consider in this work only *oriented* (or directed) goals $s \rightarrow^? t$ with solutions θ such that $\theta s \stackrel{*}{\longrightarrow} t$. Systems of such goals are used directly for lazy narrowing. For plain narrowing, it suffices to consider narrowing derivations starting from one term, here s.

In other works, solutions with reduction in both directions, i.e. $\theta s {\downarrow} \theta t$, are considered. Directed goals simplify the technical treatment in many respects and are essential for some refinements. For instance, we have shown in Section 6.3.1 that strong invariants for sets of directed goals are possible for functional-logic programming and permit deterministic variable elimination. Directed goals are also more appropriate for programming language applications, as they are operationally more perspicuous. The expressiveness lost by this assumption can easily be recovered for confluent rewrite rules by the following technique: add an equality predicate $=^?$ and the rule $X =^? X \rightarrow \mathit{true}$ to a rewrite system R. Then the R-unification problem of two terms s and t can be stated as $s = t \rightarrow^? \mathit{true}$ and solved by narrowing. This yields a semi-decision procedure for unification modulo a convergent R, as narrowing is complete w.r.t. normalized substitutions. It is important to observe that this added rule $X =^? X \rightarrow \mathit{true}$ does not destroy convergence. Notice that this rule is not left-linear, which is essential for some refinements regarding programming. In essence, this shows that for left-linear rules, there is a difference between matching and unification. For instance, there are cases where matching is decidable but unification is not [DMS92, Pre94c].

Chapter 7

Variations of Higher-Order Narrowing

This chapter discusses alternative approaches for solving higher-order equations by narrowing. Most of them are inspired by the different notions of first-order narrowing. Compared to lazy narrowing, for all of them new problems arise due to the higher-order case. For an overview of the approaches, we refer again to Figure 2.2.

The first approach we consider is the general notion of (plain) narrowing, for which many refinements exist, e.g. basic narrowing [Hul80]. The idea of this approach is to find an instance of a term such that a rewrite step somewhere in the term becomes possible. For this, Section 7.1 presents an abstract view of higher-order narrowing, where a problem with locally bound variables in the solutions becomes apparent. We show in Section 7.2 that the first-order notion of plain narrowing can be lifted to higher-order patterns and argue that it is problematic when going beyond higher-order patterns. In contrast, in the general approach in Section 7.1 most real problems are hidden in the unification. We discuss some of these in Section 7.3.

Another approach to higher-order narrowing is discussed in the last section of this chapter. The difficulties of plain narrowing in the higher-order case come from the fact that narrowing at variable positions is needed. Section 7.4 shows that we can factor out this complicated case by flattening the terms to patterns plus adding some constraints. Then narrowing on the pattern part proceeds almost as in the first-order case and it remains to solve the constraints, which can be done by lazy narrowing. In that way we have a modular structure, and higher-order lazy narrowing is used only where needed.

7.1 A General Notion of Higher-Order Narrowing

The idea of first-order plain narrowing is, roughly speaking, to find an instance of a term such that some subterm can be rewritten. Repeating this yields a complete method for matching modulo a theory given by a convergent rewrite system R, which easily extends to equational unification. We show in the following that this idea cannot be lifted directly to the higher-order case. The problem is the notion of subterm. We show that the syntactic notion of the first-order case has to be generalized, which results in highly unrestricted narrowing rules. In the two versions we present, the syntactic guidance of the starting term, as in the first-order case, is lost.

Since λ-calculus can express a notion of subterm, we can model narrowing abstractly in our setting. We simulate a context where reduction takes place by an appropriate higher-order variable C, i.e. instead of $s \longrightarrow^{l \to r} t$ we can write $s = \theta C(l) \longrightarrow \theta C(r) = t$ for an appropriate substitution θ. For instance, to rewrite $c(f(X))$ with $f(X) \to g(X)$, it suffices to take $C \mapsto \lambda x.c(x)$. This yields the following generalization of first-order narrowing, where most of the real problems are hidden in the unification.

Already in this very general setting we will identify a problem with locally bound variables in solutions. To handle bound variables correctly within λ-calculus, it will be necessary to lift l in $C(l)$ over an arbitrary set of variables, which is clearly unsatisfactory. This is shown in the following definition:

Definition 7.1.1 A term s **narrows** to t with a rule $l \to r$ and with a substitution θ, written as $s \leadsto_{\theta}^{l \to r} t$, if

- τ is a $\overline{y_k}$-lifter of l,
- θ is a unifier of $s =^? C(\lambda \overline{y_k}.\tau l)$, where C is a new variable of appropriate type, and
- $t = \theta C(\lambda \overline{y_k}.\tau r)$.

A few comments are in order:

- The $\overline{y_k}$-lifter employed is completely arbitrary, any $k \geq 0$ is possible. This causes infinite branching.
- Even for restricted left-hand sides the relation may not be decidable.
- The equation $s =^? C(l)$ may be a flex-flex pair; such pairs are usually not solved, as only higher-order pre-unification is used in applications. Furthermore, for such equations, minimal complete set of unifiers may not exist.

- Instead of explicitly replacing the subterm at position p, we use β-reduction for this purpose. It is possible to make the subterm explicit where the replacement takes place, but this considerably complicates the completeness proof.

- Note that l may occur repeatedly or not at all in $\theta C(l)$, i.e. $\theta s = \theta t$ is possible.

Lemma 7.1.2 (One Step Lifting) *Let R be a GHRS and let $l \to r \in R$. Suppose we have two terms s and t with $\theta s = t$ for a substitution θ and a set of variables V such that $\mathcal{FV}(s) \cup \mathcal{Dom}(\theta) \subseteq V$. If $t \longrightarrow_p^{l\to r} t'$, then there exist a term s' and substitutions δ and σ such that*

- $s \rightsquigarrow_\sigma^{l\to r} s'$,
- $\delta s' = t'$,
- $\delta\sigma =_V \theta$,
- $\mathcal{FV}(s') \cup \mathcal{Dom}(\delta) \subseteq V - \mathcal{Dom}(\sigma) \cup \mathcal{Rng}(\sigma)$.

Proof Assume $\overline{y_k} = \mathcal{BV}(t,p)$ and $t \longrightarrow_{\tau,p}^{l\to r} t'$, where $l \to r \in R$ is a $\overline{y_k}$-lifted rule, away from V. Let $\delta' = \theta \cup \{C \mapsto \lambda x.t[x(\overline{y_k})]_p\} \cup \tau$. Then δ' is a unifier of $s =^? C(\lambda\overline{y_k}.l)$. Let σ be a more general unifier of δ' such that $\delta' =_{V'} \delta\sigma$ for some δ and $V' = V \cup \{C\} \cup \mathcal{FV}(l)$. Assume (w.l.o.g.) $\mathcal{Dom}(\delta) \subseteq \mathcal{FV}(\sigma s)$. As $\delta' =_V \theta$, we have $\theta =_V \delta\sigma$. Then, by definition, $s \rightsquigarrow_\sigma^{l\to r} s'$ and $\delta s' = t'$ follows from

$$\delta s' = \delta\sigma C(\lambda\overline{y_k}.r) = \delta' C(\lambda\overline{y_k}.r) = t[(\lambda\overline{y_k}.\delta' r)\overline{y_k}]_p = t'.$$

Using $\mathcal{FV}(r) \subseteq \mathcal{FV}(l)$ we obtain

$$\begin{aligned} \mathcal{FV}(s') &= \mathcal{FV}(\sigma C(\lambda\overline{y_k}.r)) \\ &\subseteq \mathcal{FV}(\sigma C(\lambda\overline{y_k}.l)) \\ &= \mathcal{FV}(\sigma s) \\ &\subseteq V - \mathcal{Dom}(\sigma) \cup \mathcal{Rng}(\sigma). \end{aligned}$$

Hence we have

$$\mathcal{FV}(s') \cup \mathcal{Dom}(\delta) \subseteq V - \mathcal{Dom}(\sigma) \cup \mathcal{Rng}(\sigma)$$

as $\mathcal{Dom}(\delta) \subseteq \mathcal{FV}(\sigma s)$. □

With Lemma 7.1.2, completeness of narrowing can be shown easily, as for instance in the next section. Notice that the above proof uses some unifier that is more general than the substitutions of the reduction considered, although it would be sufficient to use the solution θ as the unifier of $s =^? C(\lambda\overline{y_k}.\tau l)$. It would be desirable to use a maximally general unifier instead, but these may not exist for higher-order unification.

For the proof of the above lemma it is important that the rewrite rule $l \to r$ has been lifted over the right number of bound variables. Let us see by an example that the number of variables over which a rule has to be lifted cannot be determined beforehand. The problem occurs when a solution θ for a variable X contains a local λy and a rewrite step in a subterm below where y occurs has to be lifted. For a narrowing step, the replaced subterm is made explicit in $\sigma C(l) \longrightarrow \sigma C(r)$, but y is not visible yet. With the lifting of $l \to r$ it is possible to rename bound variables in r later. A somewhat similar problem with higher-order matching was reported in [Pau86] and [PE88].

Example 7.1.3 Assume $R = \{h(P,a) \to g(P,a)\}$ and consider the equational matching problem

$$H(a) \to^? u(\lambda y.g(y,a))$$

with the solution $\{H \mapsto \lambda x.u(\lambda y.h(y,x))\}$. When narrowing without lifting, we obtain $H(a) \leadsto^R H''(g(P',a))$, which matches $u(\lambda y.g(y,a))$, but does not subsume the above solution, as $g(P',a)$ cannot be instantiated to $g(y,a)$.

The solution is obtained here by lifting the rule over one parameter. First, the solution to the unification problem

$$H(a) =^? C(\lambda y.h(P(y),a)),$$

which is needed for the narrowing step, is

$$\{H \mapsto \lambda x.H'(\lambda y.h(P(y),x)), C \mapsto \lambda x.H'(\lambda y.x(y))\}.$$

Then we have $H(a) \leadsto^R H'(\lambda y.g(P(y),a))$ and the matching problem can be solved with the substitution $\{H' \mapsto \lambda x.u(x), P \mapsto \lambda x.x\}$. In the general case, the solution to H may contain an arbitrary number of locally bound variables, such as y here, but the need to lift over these variables is not visible when looking at $H(a)$. To obtain completeness for this definition of narrowing, we thus have to guess locally bound variables, at least in our framework.

The above notion of narrowing is not of great computational interest. For instance, there is little hope of finding cases where even the application of narrowing is decidable. The purpose of the above was to show that the first-order narrowing idea of unifying a subterm with a left hand side cannot be lifted to the higher-order case.

Another, somewhat similar approach to mention here is the notion of higher-order rewriting by van de Pol in [Pol94]: For a rewrite rule $l \to r$, we abstract over all free variables $\overline{x_k}$ in $\lambda\overline{x_k}.l$. Then a rewrite step from s to t can be defined via

$$s = \theta C(\lambda\overline{x_k}.l) \longrightarrow \theta C(\lambda\overline{x_k}.r) = t$$

This definition was motivated by some results on termination and is clearly far removed from the first-order notion of narrowing at a subterm. It seems possible to model higher-order narrowing by this approach without guessing arbitrary bound variables. We simply have to replace matching of a term as in a rewrite step by unification. For instance, consider the above example with the adapted unification problem

$$H(a) =^? C(\lambda p.h(p,a))$$

This can be solved with

$$\{H \mapsto \lambda x.H'(\lambda p.h(p,x)), C \mapsto \lambda x.H'(\lambda y.x(y))\}.$$

The corresponding narrowing step yields the goal

$$H'(\lambda p.g(p,a)) \to^? u(\lambda y.g(y,a))$$

which can be solved similarly as in the above.

Although this alternative may have some advantages, the notion of subterm is also too general to be of any practical use. It does not solve the main problem, which is the degree of non-determinism. Particularly, solving flex-flex pairs is needed.

In summary, there is little hope of obtaining a definition with similar properties as in the first-order case. The only positive subcase are patterns, for which we show in the following section that the first-order notion of narrowing can be adapted.

7.2 Narrowing on Patterns with Pattern Rules

In this section we show that the first-order notion of (plain) narrowing can be adapted to a restricted set of λ-terms, higher-order patterns. Then, as in the first-order case, narrowing at variable positions implies that the used substitution is reducible, thus this step is redundant.

Assumption. We assume in this section that all terms, including the rewrite rules, are patterns.

Although pattern rules are not sufficient for expressing higher-order functional programs (see e.g. Section 9.1), there are examples from other areas, where bound variables are involved. For instance, scoping rules for quantifiers (as in [Nip91a]), e.g.

$$P \wedge \forall x.Q = \forall x.(P \wedge Q),$$

can be expressed by patterns.

Definition 7.2.1 A **pattern narrowing** step from a pattern s to t with a pattern rule $l \longrightarrow r$ at a non-variable position q with substitution θ is defined as $s \underset{p}{\leadsto}{}^{l\to r}_{q,\theta}\ t$, where

- τ is a $\overline{y_k}$-lifter of l, where $\overline{y_k} = \mathcal{BV}(s,q)$,
- θ is a most general unifier of $\lambda\overline{y_k}.s|_q$ and $\lambda\overline{y_k}.\tau l$, and
- $t = \theta(s[\tau r]_q)$.

This notion of narrowing coincides with the standard definition of first-order narrowing on first-order terms. Here, in contrast to the notion of narrowing in Section 7.1, we only have to lift the rule $l \longrightarrow r$ into the context at position q. The problem in Section 7.1 with locally bound variables occurs only when narrowing at variable positions, which is not needed here. When working with first-order equations, as done by Qian [Qia94] and by Snyder [Sny90], this lifting is not strictly needed, as the bound variables in $s|_q$ can be treated as new constants and/or ignored. For rewriting higher-order patterns with first-order rules, the bound variables play no role, as they cannot occur in the part of the term which matches the rule. In these lines, the results by Qian lift completeness of first-order narrowing strategies to patterns for first-order equations. We conjecture that most first-order narrowing strategies can also be lifted to our setting, yet not as in [Qia94].

For a sequence

$$s_0 \underset{p}{\leadsto}{}^{R}_{\theta_1} s_1 \underset{p}{\leadsto}{}^{R}_{\theta_2} \cdots \underset{p}{\leadsto}{}^{R}_{\theta_n} s_n$$

we write $s_0 \underset{p}{\overset{*}{\leadsto}}{}^{R}_{\theta} s_n$, where $\theta = \theta_n \ldots \theta_1$. We first lift one rewrite step in a solution to one narrowing step. The lemma and its proof resemble closely their first-order counterparts, as e.g. in [MH94, Hul80]. This result has been developed independently by the author in [Pre94b] and in [ALS94a, LS93] for conditional rules.[1]

[1] See Section 9.1 for more details.

Lemma 7.2.2 (One Step Lifting) *Let R be a pattern HRS and let $l \to r \in R$. Suppose we have two patterns s and t with $t = \theta s$ for an R-normalized substitution θ, and a set of variables V such that $\mathcal{FV}(s) \cup \mathcal{Dom}(\theta) \subseteq V$. If $t \longrightarrow^{l \to r} t'$, then there exist a term s' and substitutions δ, σ such that*

- $s \underset{p}{\leadsto}{}^{l \to r}_{\sigma} s'$,
- $\delta s' = t'$,
- $\delta\sigma =_V \theta$,
- *δ is R-normalized, and*
- $\mathcal{FV}(s') \cup \mathcal{Dom}(\delta) \subseteq V - \mathcal{Dom}(\sigma) \cup \mathcal{Rng}(\sigma)$.

Proof Assume $\theta s \longrightarrow^{l \to r}_{p,\varphi} t'$ and $l' \to r'$ is a rule lifted over $\overline{y_k}$ from $l \to r$ away from V. As l is of base type, $\theta s|_p$ cannot be an abstraction. We have $(\theta s)|_p = \varphi l'$. Since θ is R-normalized, p is a non-variable position in s and $\theta s|_p = \theta(s|_p)$.

Let σ be a most general unifier of $\lambda \overline{y_k}.s|_p$ and $\lambda \overline{y_k}.l'$ such that there exists δ with $\delta\sigma =_V \theta$. Assume that δ is minimal, i.e. $\mathcal{Dom}(\delta) \subseteq V - \mathcal{Dom}(\sigma) \cup \mathcal{Rng}(\sigma)$ holds. Since σ is a pattern substitution, δ is R-normalized, as θ is.

Then, by definition, $s \underset{p}{\leadsto}{}^{l \to r}_{\sigma,p} s' = \sigma s[r']_p$. To see that this step lifts the rewrite step on θs, it remains to show

$$\delta s' = \delta\sigma s[r']_p = \theta(s[r']_p) = t'.$$

The second equation follows from $\delta\sigma =_V \theta$. Then from $\mathcal{FV}(r) \subseteq \mathcal{FV}(l)$ and from $\mathcal{FV}(\sigma s|_p) \subseteq \mathcal{FV}(\sigma l)$

$$\mathcal{FV}(s') \subseteq \mathcal{FV}(\sigma s, \sigma r) \subseteq \mathcal{FV}(\sigma s, \sigma l) \subseteq \mathcal{FV}(\sigma s) \subseteq V - \mathcal{Dom}(\sigma) \cup \mathcal{Rng}(\sigma)$$

follows. Hence we have $\mathcal{FV}(s') \cup \mathcal{Dom}(\delta) \subseteq V - \mathcal{Dom}(\sigma) \cup \mathcal{Rng}(\sigma)$ as δ is minimal, which concludes the proof. □

The following lemma holds for patterns as for first-order terms [MH94].

Lemma 7.2.3 *Let σ, θ, θ' be pattern substitutions and V, V' be sets of variables such that $(V' - \mathcal{Dom}(\sigma)) \cup \mathcal{Rng}(\sigma) \subseteq V$. If $\theta =_V \theta'$ then $\theta\sigma =_{V'} \theta'\sigma$.*

Completeness of narrowing follows as in the first-order case:

Theorem 7.2.4 (Completeness of Pattern Narrowing) *Let $\mathcal{R}$ be a pattern HRS. Suppose we have terms s and $t = \theta s$ for a R-normalized substitution θ and a set of variables V such that $\mathcal{FV}(s) \cup \mathcal{Dom}(\theta) \subseteq V$. If $t \stackrel{*}{\longrightarrow}{}^R t'$, then there exist a term s' and substitutions δ, σ such that*

- $s \overset{*}{\underset{p}{\rightsquigarrow}}{}^{R}_{\sigma}\, s'$,
- $\delta s' = t'$,
- $\delta\sigma =_V \theta$, *and*
- $\mathcal{FV}(s') \cup \mathcal{D}om(\delta) \subseteq V - \mathcal{D}om(\sigma) \cup \mathcal{R}ng(\sigma)$.

Proof by induction on the length of the reduction from t to t'. Assume $t \longrightarrow^{l \to r}_{\varphi} t_1$. By Lemma 7.2.2 there exist a term s_1 and substitutions δ_1, σ_1 such that

- $s \underset{p}{\rightsquigarrow}{}^{l \to r}_{\sigma_1}\, s'$,
- $\delta_1 s' = t_1$,
- $\delta_1 \sigma_1 =_V \theta$,
- δ_1 is R-normalized, and
- $\mathcal{FV}(s') \cup \mathcal{D}om(\delta_1) \subseteq V - \mathcal{D}om(\sigma_1) \cup \mathcal{R}ng(\sigma_1)$.

Let $V_1 = V - \mathcal{D}om(\sigma_1) \cup \mathcal{R}ng(\sigma_1)$. Then the induction hypothesis yields

- $s_1 \overset{*}{\underset{p}{\rightsquigarrow}}{}^{R}_{\sigma_2}\, s'$,
- $\delta s_1' = t'$,
- $\delta \sigma_2 =_{V_1} \delta_1$,
- δ is R-normalized, and
- $\mathcal{FV}(s') \cup \mathcal{D}om(\delta) \subseteq V_1 - \mathcal{D}om(\sigma_2) \cup \mathcal{R}ng(\sigma_2)$.

Let $\sigma = \sigma_2 \sigma_1$. Then $\delta\sigma =_V \theta$ follows as $\delta\sigma_2 =_{V_1} \delta_1$ and $\delta\sigma_2\sigma_1 =_V \delta_1\sigma_1$ yields $\delta\sigma_2\sigma_1 = \delta\sigma$ with Lemma 7.2.3 and hence $\delta\sigma = \theta$. □

7.3 Narrowing Beyond Patterns

We discuss in the following some of the difficulties which occur when extending the first-order notion of plain narrowing for patterns to full λ-terms. This will show that it is difficult to find a sufficiently powerful notion of narrowing in the lines of pattern narrowing, which is more practical than the one in Section 7.1. Thus the section does not present a new approach, but more a justification for choosing different approaches.

We use in this section the relation $\rightsquigarrow$ in a more informal way to exemplify the problems involved. In the following, we allow λ-terms in both the rules as in the goals. In Example 7.1.3 we have identified a problem with locally bound variables. This and several other problems stem from the fact that narrowing at variable positions is required, since the rewrite step we lift might have been at a reducible subterm created by β-reduction. We discuss this with the following example.

Example 7.3.1 Assuming the rewrite system

$$R_0 = \{f(f(X)) \rightarrow g(X)\},$$

narrowing at a variable position is required to find the solution $\{H \mapsto \lambda x.f(x)\}$ to the problem $\lambda x.H(f(x)) \rightarrow^? \lambda x.g(x)$:

$$\lambda x.H(f(x)) \rightsquigarrow^{R_0}_{H \mapsto \lambda x.f(x)} \lambda x.g(x)$$

Now the problem is how to define narrowing at variable positions. For instance, consider the solution $\theta = \{H \mapsto \lambda x.h(f(x),x)\}$ to the equational problem

$$\lambda x.H(f(x)) \rightarrow^? \lambda x.h(g(x),f(x)),$$

w.r.t. the R_0-reduction

$$\lambda x.h(f(f(x)),f(x)) \longrightarrow^{R_0} \lambda x.h(g(x),f(x)).$$

The naive approach, to instantiate H as little as possible, as in

$$\lambda x.H(f(x)) \rightsquigarrow^{R_0}_{H \mapsto \lambda x.H'(f(x))} \lambda x.H'(g(x)),$$

fails. The problem is that the subterm $f(x)$ is duplicated by θ and the reduction does not occur inside $f(x)$. An idea is to create a "local context" at this variable. Hence, we instantiate H first with $\{H \mapsto \lambda x.H''(H'(x),x)\}$. Then, after β-reduction, the subterm $H'(f(x))$ can be unified by $\{H' \mapsto \lambda x.f(x)\}$ with the left-hand side $f(f(x))$ and can be rewritten. Thus we have

$$\lambda x.H(f(x)) \rightsquigarrow^{R_0}_{H \mapsto \lambda x.H''(f(x),x)} \lambda x,y.H'(g(x),f(x))$$

and the solution, here $\{H'' \mapsto \lambda x,y.h(x,y)\}$, is then obtained by unification.

Intuitively, we approximate the desired solution θH in the first argument of H''. A further problem occurs when narrowing on an argument of a free variable. For instance, assume the narrowing step

$$H(f(X)) \rightsquigarrow^{R_0}_{X \mapsto f(Y)} H(g(Y)).$$

Then some solution to H may copy the argument of H, thus this narrowing step corresponds to several rewrite steps. As a consequence, the solution $\{X \mapsto f(Y), H \mapsto \lambda x.h(x,x)\}$ with the reduction

$$H(f(f(Y))) \rightarrow^{?} h(g(Y), f(f(Y)))$$

to the above matching problem will not be found with the narrowing step above. The reducible subterm is copied in the solution, but for narrowing, only one copy is visible.

Due to all these problems, we do not develop the notion of plain narrowing further. This explains our focus on the alternative, lazy narrowing, in the previous chapter. In addition, we develop another approach that extends pattern narrowing by additional higher-order constraints in the following.

7.4 Narrowing on Patterns with Constraints

We have seen in Section 7.1 that the well-developed first-order notion of plain narrowing is problematic when going beyond higher-order patterns. Although lazy narrowing solves most of these problems, it would be nice to integrate some of the ideas of the former approach.

An approach that allows the use of plain narrowing in the higher-order case is presented in this section. The idea is to factor out the complicated case, narrowing at variable positions, into constraints and to work with the simpler pattern part as shown in Section 7.2. The idea is similar to [Pfe91], where non-pattern unification problems are delayed in a higher-order logic programming language. In contrast to the latter, we also have to solve the constraints modulo R.

The rules NC in Figure 7.1 work on a pair (t, C), where t is a goal, in which non-pattern subterms can be shifted to the goals C with rule Flatten. These can be solved with lazy narrowing as done here, or any comparable method. Then for the term t, narrowing at or below variable positions is not needed. The assumption is that in many applications, most (sub-)terms are patterns, such that the pattern part performs the large part of the computation.

For instance, to solve a goal $f(F(f(a))) \rightarrow^{?} g(a)$ w.r.t. R_0 as in Example 7.3.1, we flatten the left-hand side to $(f(F') \rightarrow^{?} g(a), \{F(f(a)) \rightarrow^{?} F'\})$. Then the flattened term can be handled with first-order techniques, possibly yielding $\{F' \mapsto f(a)\}$. Solving the remaining constraint $F(f(a)) \rightarrow^{?} f(a)$ is simple, and it may not even be desirable to compute all its solutions.

The rule Pattern Narrow applies only at subterms that have been flattened to patterns. Hence the unification needed in this rule is pattern unification.

Solve

$$(t \to^? t', C) \;\Rightarrow^\theta\; (t' \to^? t', \theta C) \text{ if } \theta t = t'$$

Flatten

$$(t \to^? t', C) \;\Rightarrow\; (t[X'(\overline{x_k})]_p \to^? t', \{\lambda \overline{x_k}.X(\overline{t_n}) \to^? X'\} \cup C)$$
if p is a rigid path in t such that $t|_p = X(\overline{t_n})$ is not a pattern, where $\overline{x_k} = \mathcal{BV}(t,p)$

Pattern Narrow

$$(t \to^? t', C) \;\Rightarrow^\theta\; (s \to^? t', \theta C) \text{ if } p \text{ is a rigid path in } t,$$
$t|_p$ is a pattern, and $t \leadsto^R_{p,\theta} s$

Constraint Solving

$$(t \to^? t', C) \;\Rightarrow^\theta\; (\theta t \to^? \theta t', C') \text{ if } C \Rightarrow^\theta_{LN} C'$$

Figure 7.1: System NC for Narrowing with Constraints

The main advantage of this version of narrowing is that we achieve a system, where we can work similar to the first-order case on the pattern part.

To prove completeness we first need a more technical lemma. The problem is that the two methods integrated here are based on very different proof strategies. The next lemma shows more precisely which rewrite steps are handled by lazy narrowing in the constraints and which are modeled directly. When working with NC we will call the goal t in a tuple (t,C) the pattern part, although it may not be a pattern goal.

Lemma 7.4.1 *Assume a convergent HRS R, two terms s and t, where t is a ground R-normal form, an R-normalized substitution* θ*, and a set of constraints* $\overline{G_n} = \{u_n \rightarrow^? u'_n\}$ *such that*

- $\theta s \stackrel{*}{\longrightarrow}^R t$ *and*
- $\overline{\theta u_n \stackrel{*}{\longrightarrow}^R \theta u'_n}$.

Then $(s \rightarrow^? t, \{\overline{G_n}\}) \stackrel{*}{\Rightarrow}^{\delta}_{NC} (s' \rightarrow^? t, \{\overline{G'_m}\})$ *such that there exists* θ' *with*

- $\theta s' = t$,
- $\theta =_{\mathcal{FV}(s)} \theta'\delta$,
- $\theta s \stackrel{*}{\longrightarrow}^R \theta' s'$,
- θ' *is R-normalized, and*
- θ' *is a solution of* $\overline{G'_m}$.

Proof by induction on $\longrightarrow^R$ on θs, which is terminating. We maintain the last four claims as invariants. Assume $\theta s \longrightarrow_p^{l\rightarrow r} t_1 \stackrel{*}{\longrightarrow}^R t$ is an innermost reduction with some appropriately lifted rule $l \rightarrow r \in R$. We have the following two cases depending on p. Since θ is normalized, p cannot occur below a pattern subterm $X(\overline{y_m})$ in s.

If p is not a position on a rigid path in s, the reduction is modeled in the constraints: let q be a minimal prefix of p such that $s|_q$ is of the form $X(\overline{t_n})$. Let $\overline{x_k} = \mathcal{BV}(t,q)$. Since θ is R-normalized, $\lambda\overline{x_k}.X(\overline{t_n})$ cannot be a pattern. Apply flatten to obtain a new constraint $G_0 = \lambda\overline{x_k}.X(\overline{t_n}) \rightarrow^? \lambda\overline{x_k}.X'(\overline{y_n})$ for a new variable X'. Let further $s' = s[X'(\overline{y_n})]q$ and $\theta' = \theta \cup \{X' \mapsto \lambda\overline{x_k}.(\theta s|_q)\downarrow_R\}$. As R is convergent, we obtain $\theta' s' \stackrel{*}{\longrightarrow} t$ via an innermost reduction. Since $\theta s \stackrel{*}{\longrightarrow}^R \theta' s'$, the induction hypothesis applies with $(s', G_0 \cup \overline{G_n})$ and θ'.

In case p is on a rigid path in s, we apply Flatten at all (maximal) non-pattern subterms $\overline{s_m}$ of $s|_p$. This yields s', some new constraints

$$\overline{G'_m} = \overline{t_m \rightarrow^? X_m}$$

and a new associated solution θ_1. As described in the last case, to obtain θ_1, θ has to be extended at each Flattening step. Since the reduction is innermost, all $\overline{\theta s_m}$ are in R-normal form and hence the new solutions added for $\overline{X_m}$ are R-normalized. Thus we have $\theta s|_p = \theta_1 s'|_p$.

Then the rule Pattern Narrow applies and the proof proceeds similar to Theorem 7.2.2: as $s_1|_p$ is a pattern and $\theta_1 s_1|_p$ is an instance of l, there exists a most general unifier δ of $s_1|_p$ and l and there exits θ' such that $\theta_1 = \theta'\delta$. Then the Pattern Narrow step yields $(s' \rightarrow^? t, \{\overline{\delta G'_m}, \overline{\delta G_n}\})$, where $s' = \delta s_1[r]_p$. It follows as in Theorem 7.2.2 that θ' is R-normalized. Clearly, θ' is a solution for all constraints $\overline{\delta G'_m}, \overline{\delta G_n}$. As $\theta s \xrightarrow{*}^R \theta_1 s_1 \xrightarrow{*}^R \theta' s'$, it remains to apply the induction hypothesis with θ' and $(s' \rightarrow^? t, \{\overline{\delta G'_m}, \overline{\delta G_n}\})$. □

Now the completeness of NC follows easily. We only have to lift rewrite steps that occur in the primary goal, the others are handled by lazy narrowing in the constraints.

Theorem 7.4.2 (Completeness of NC) *Assume a convergent HRS R. If $s \rightarrow^? t$ has the solution $\theta s \xrightarrow{*}^R t$ where θ is R-normalized and t is a ground R-normal form, then $(s \rightarrow^? t, \{\}) \stackrel{*}{\Rightarrow}{}^{\delta}_{NC} (t \rightarrow^? t, C)$ such that $\delta s = t$ and δ is more general modulo the newly added variables than θ and the goals in C are flex-flex.*

Proof First, apply Lemma 7.4.1 to $(s \rightarrow^? t, \{\})$, yielding a pair $(s' \rightarrow^? t, \{\overline{G_n}\})$ which is solvable by some substitution θ' with $\theta \leq \theta'$. Thus the Solve rule applies with some substitution $\delta \leq \theta'$. It remains to solve the constraints $\{\overline{\delta G_n}\})$ by System LN. □

It is interesting to examine how rewrite steps in a solution $\theta s \longrightarrow^R s_1 \xrightarrow{*}^R t$ are modeled in the pattern part in above completeness result. There are two possibilities for a rewrite step $\theta s \longrightarrow^R s_1$ at a position p:

- If there exists a prefix q of p such that $s|_q$ is a non-pattern term, then this non-pattern subterm is flattened into the constraints and replaced by a new variable, say X. The solution associated to this (intermediate) variable X is the R-normal form of $s|_q$, thus the flattening step shifts the normalization of a full subterm into the constraints.

- Otherwise, the subterm is flattened to a pattern. Then a single narrowing step is lifted and this step takes place at the same position and with the same rule as the narrowing step. As in the first-order case, we have a one-to-one correspondence of the rewrite step in θs and the narrowing step in s.

With the last observation in mind, we conjecture that narrowing strategies for first-order rewrite systems can be lifted to the pattern part in a modular way. The problem is that most first-order strategies only lift particular derivations, e.g. innermost reductions (basic narrowing [Hul80]) or leftmost innermost reductions (LSE narrowing [BKW93]). A reduction is leftmost, if for each step no rewrite step at a position left to it applies. However, for leftmost innermost solutions $\theta s \xrightarrow{*}^R t$, it seems that the above completeness result can be extended to show that the reductions in the pattern part form a leftmost innermost subsequence of the $\theta s \xrightarrow{*}^R t$ reduction.

Chapter 8

Applications of Higher-Order Narrowing

This section presents examples for higher-order rewriting and narrowing. Most of these applications fall into functional-logic programming, for which left-linear rewrite rules and thus Simple Systems suffice. Only the examples in Section 8.2 on program transformation and type inference go beyond programming with left-linear rules and more expressiveness is useful. For other examples on the utility of higher-order constructs, we refer to [Nip91a, MN97] for formalizing logics and λ-calculi, and for process algebras to [Pol94]. Other application areas are program synthesis [Hag91b], machine learning [Har90, DW88], natural language processing [Nad87, PM90, GK96, GKL96], and theorem proving systems, for instance see [AINP90, Pau94, Gor88, CAB$^+$86, DFH$^+$93].

Our higher-order setting allows for several new aspects in functional-logic programming, some of which are summarized in the following:

- Higher-order functional programming with search is useful in several examples. For instance, the parsing example integrates ideas of logic and higher-order functional programming.

- $\forall$-quantified goals can be modeled directly via binders. That is, a goal $\forall x.(t \to^? s)$ is logically equivalent to $\lambda x.t \to^? \lambda x.s$. This extension of logic programming will be used in several of the following examples. We prefer the notation $\forall x$ instead of λx to clarify the examples. In this way, bound variables permit precise control on variables depending on a quantified variable. This is similar to lifting over parameters. For instance, in a goal $\forall x.f(F(x),x)$, the term $F(x)$ depends on x, and thus the goal is different from $\forall x.f(F,x)$. More formally,

  - $\exists y.\forall x.p(x,y)$ is modeled as $\lambda x.p(x,Y)$ and
  - $\forall x.\exists y.p(x,y)$ is modeled as $\lambda x.p(x,Y(x))$.

  In the general case, we may have so-called mixed prefixes (of existential and universal quantifiers), which are treated in depth in [Mil92].

- λ-terms are often useful as data structures, as λ-calculus provides for built-in anonymous local constants via α-conversion and a notion of substitution via β-reduction. This is similar to higher-order logic programming.

  These aspects of λ-terms are often essential, as for instance in the *diff*-example. In several other domains, such as dealing with programs, e.g. modeling semantics of programs and hardware synthesis, the constructs and conversions of λ-calculus are natural.

- In contrast to the first-order case, it is possible to compute or synthesize functions via unification. This is even needed if the original goal does not contain higher-order free variables, since such variables may be introduced during computation.

- Partial evaluation of functional programs by rewriting is the essence of many examples and is in particular often used for $\forall$-quantified goals. Note that this is not possible in functional languages.

- Functional objects can be treated as data objects. This means that functions have a term structure which can be accessed. For instance, for a functional term $\lambda x.and(x,or(a,x))$, we can write a program to count the number of function symbols *and* and *or*. This restricts the computed functional objects to consist only of certain function symbols or to be of a certain maximal size. (See the example on hardware synthesis in Section 8.1.1.)

8.1 Functional-Logic Programming

Our approach to functional-logic programming is oriented towards functional languages. Thus we show in the following how to model some basic programming schemes in a functional style. This differs from many first-order approaches which often aim at extending Prolog by functions. Our goal is to extend a functional core language by logical variables as in Prolog. Relational programming as in logic programming can be embedded as shown in Section 2.8.2.

We show by several examples that left-linear, normal conditional HRS suffice for programming. This allows computation with Simple Systems. Since we do not allow extra variables on the right-hand side of the conditions, local variables as in functional programming are created via let-constructs, as for instance shown in Section 2.8.2. For example, we show how the let-construct for pairs used in Section 2.8.2 can be formulated by higher-order rewrite rules. This common notation for let can be desugared by

$$\mathsf{let}\ pair(xs,ys) = X\ \mathsf{in}\ F(xs,ys) =^{def} let\ X\ in\ \lambda xs,ys.F(xs,ys).$$

Notice that in the sugared notation on the left, $pair(xs,ys)$ serves as a binder for xs and ys. The higher-order rewrite rule for this construct is

$$let\ pair(Xs,Ys)\ in\ \lambda xs,ys.F(xs,ys) \to F(Xs,Ys).$$

The idea behind this modeling is that in $let\ t\ in\ \lambda xs,ys.t'$, the term t is evaluated to a constructor rooted term of the form $pair(Y,Z)$ and then the rewrite rule applies.

Several of the following examples assume an equality predicate on natural numbers. There are two ways to formalize such a predicate: either simply by a rule $X =^? X \to true$, which goes beyond Simple Systems, or by encoding strict equality $=_s$ on numbers, as shown in Section 6.5.1. As we will see, strict equality suffices for most applications. Recall that we sometimes write p for a rule $p \to true$ or a goal $p \to^? true$. Furthermore, we use in the examples some common abbreviations, e.g. $1 = s(0)$ etc.

8.1.1 Hardware Synthesis

This section shows how simple hardware gates can be represented and computed within functional-logic programming. Representing circuits by λ-terms is a very natural and convenient choice. This allows simple composition of circuits and circuit computation, which is just application plus evaluation. At the same time, it is possible to access the term structure of the circuit. This will be used below to restrict the search space. Another advantage of this representation is that equality of circuits is irrespective of the names of input parameters.

The goal of this example is to compute functions composed of *nand*-functions (or gates). Thus we first specify the *nand*-function and some auxiliary functions, as shown in Figure 8.1. In this set of rules, the function *size_nand* serves two purposes. First, it counts the number of *nand* functions, but also assures that some term contains no other functions.

$$
\begin{array}{lcl}
nand(0,X) & \to & 1 \\
nand(X,0) & \to & 1 \\
nand(1,1) & \to & 0 \\
map2(F,[]) & \to & [] \\
map2(F,[pair(X,Y)|R]) & \to & [F(X,Y)|map2(F,R)] \\
size\_nand(\lambda x,y.x) & \to & 0 \\
size\_nand(\lambda x,y.y) & \to & 0 \\
size\_nand(& & \\
\quad \lambda x,y.nand(F(x,y),G(x,y))) & \to & size\_nand(\lambda x,y.F(x,y))+ \\
 & & size\_nand(\lambda x,y.G(x,y))
\end{array}
$$

Figure 8.1: Rules for Hardware Synthesis

Now we are ready to compute a few simple gates. We synthesize an *or*-function consisting of at most three *nand*-gates with the following goals

$$
\begin{array}{l}
size\_nand(F) \leq 3 \to^? true, \\
\forall x,y.map2(F,[pair(0,0),pair(x,1),pair(1,y)]) \to^? [0,1,1],
\end{array}
$$

where the second goal specifies the *or*-function. The first solution

$$F \mapsto \lambda x,y.nand(nand(x,x),nand(y,y))$$

is found by exhaustive search. The only other solution is a simple permutation:

$$F \mapsto \lambda x,y.nand(nand(y,y),nand(x,x))$$

It is also possible to compute functions with one argument by ignoring the other. For instance, a *not*-function consisting of at most two *nand*-gates can be specified with

$$
\begin{array}{lcl}
size\_nand(\lambda x,y.F(x)) \leq 2 & \to^? & true, \\
map(F,[0,1]) & \to^? & [1,0].
\end{array}
$$

The solution is simply $F \mapsto \lambda x.nand(x,x)$.

8.1.2 Symbolic Computation: Differentiation

In this section we elaborate and summarize the example modeling symbolic differentiation. Symbolic differentiation is a standard example in many text books on Prolog [SS86]. In contrast to first-order programming, we can easily

$$
\begin{array}{lcl}
diff(\lambda y.F, X) & \to & 0 \\
diff(\lambda y.y, X) & \to & 1 \\
diff(\lambda y.sin(F(y)), X) & \to & cos(F(X)) * diff(\lambda y.F(y), X) \\
diff(\lambda y.cos(F(y)), X) & \to & -1 * sin(F(X)) * diff(\lambda y.F(y), X) \\
diff(\lambda y.F(y) + G(y), X) & \to & diff(\lambda y.F(y), X) + diff(\lambda y.G(y), X) \\
diff(\lambda y.F(y) * G(y), X) & \to & diff(\lambda y.F(y), X) * G(X) + \\
 & & diff(\lambda y.G(y), X) * F(X) \\
diff(\lambda y.ln(F(y)), X) & \to & diff(\lambda y.F(y), X)/F(X) \\
X \; * \; 1 & \to & X \\
1 \; * \; X & \to & X \\
X \; * \; 0 & \to & 0 \\
0 \; * \; X & \to & 0 \\
0 \; + \; X & \to & X \\
X \; + \; 0 & \to & X \\
0 \; / \; X & \to & 0
\end{array}
$$

Figure 8.2: Rules R_d for Symbolic Differentiation

formalize functions to be differentiated. To model a functional term $\lambda x.t$, first-order implementations use either free variables [SS86] or "special" constants for the parameter x [Bac91]. Hence first-order versions are often unsound for the general case and only work for particular goals (usually ground terms). This will be discussed in detail below.

In our setting, we define a function *diff* such that $diff(\lambda x.v, X)$ computes the value of the differential of $\lambda x.v$ at X. Note that this rule can be abbreviated by a more concise rule of non-base type: $diff(\lambda x.F) \to \lambda x.0$.

Figure 8.2 shows the rules of R_d for symbolic differentiation with left-linear, second-order equations of base type. Observe that we do not formalize the chain rule explicitly, as this would require nested free variables. As our goal is to have patterns as left-hand sides, the term $\lambda x.diff(F(G(x)))$ is not desirable as the left-hand side of a general chain rule. Furthermore, this non-pattern left-hand side gives rise to a non-terminating rewrite rule. Notice that the right-hand sides of the rules of R_d in Figure 8.2 are non-patterns, hence rewriting a pattern term may yield a non-pattern.

Termination of the rules by the method developed in [Pol94] is shown in [Pre95c]. Next we show that R_d is confluent via critical pair analysis. It is easy to see that the first rule overlaps with all remaining rules for *diff* and the remaining rules only have trivial critical pairs. All of these are joinable; for instance, the pair

$$(0, diff(\lambda x.F) * G + diff(\lambda x.G) * F),$$

generated from $diff(\lambda x.F * G, X)$, is joinable. Thus R_d is a convergent, left-linear HRS and we can apply Simple Systems with normalization.

Consider the following lazy narrowing derivation, which was shown partly in earlier chapters.

$$\begin{array}{ll}
\{\lambda x.diff(\lambda y.ln(F(y)),x) \to^? \lambda x.cos(x)/sin(x)\} & \stackrel{*}{\Rightarrow} \text{Evaluation for diff} \\
\{\lambda x.diff(\lambda y.F(y),x)/F(x) \to^? \lambda x.cos(x)/sin(x)\} & \stackrel{*}{\Rightarrow} \text{Decomposition} \\
\{\lambda x.diff(\lambda y.F(y),x) \to^? \lambda x.cos(x), & \\
\lambda x.F(x) \to^? \lambda x.sin(x)\} & \stackrel{*}{\Rightarrow} \text{Elimination} \\
\{\lambda x.diff(\lambda y.sin(y),x) \to^? \lambda x.cos(x)\} & \stackrel{*}{\Rightarrow} \text{Evaluation} \\
\{\lambda x.cos(x) * diff(\lambda y.y,x) \to^? \lambda x.cos(x)\} & \stackrel{*}{\Rightarrow} \text{Evaluation} \\
\{\lambda x.cos(x) \to^? \lambda x.cos(x)\} & \stackrel{*}{\Rightarrow} \text{Decomposition} \\
\{\} &
\end{array}$$

Due to the call-by-need strategy, there is no search necessary to find the solution $F \mapsto \lambda x.sin(x)$.

It is interesting to apply pattern narrowing to this example, although the rewrite system has non-pattern rewrite rules. The solution can be found with the pattern narrowing sequence

$$\begin{array}{ll}
\lambda x.diff(\lambda y.ln(F(y)),x) & \longrightarrow \\
\lambda x.diff(\lambda y.F(y),x)/F(x) & \rightsquigarrow \\
\lambda x.cos(F'(x)) * diff(\lambda y.F'(y),x)/sin(F'(x)) & \stackrel{*}{\rightsquigarrow} \\
\lambda x.cos(x)/sin(x), &
\end{array}$$

as all terms occurring are patterns. This derivation is shorter than the above, but the reduction is not goal directed and there are many alternative search branches. For instance, there are several alternative narrowing steps at the subterm $diff(\lambda y.F'(y),x)$ in the third term above. This clearly shows the need for further refinements.

To conclude this section, we compare our version to some first-order ones in the literature. The problem is to model the functional term in $diff(\lambda x.t)$. By some authors, e.g. [Bac91], the parameter x is modeled as a "special constant" x with rules of the form:

$$\begin{array}{rcl}
diff'(x) & \to & 1 \\
diff'(0) & \to & 0 \\
& \vdots &
\end{array}$$

This is quite limited, since only differentiation in x is possible and x may not occur elsewhere. As there are no bound variables, a rule (or goal) with $\mathit{diff}(\lambda x.F, X)$ on the left is not possible. Hence a goal $\mathit{diff}'(F) \rightarrow^? 0$ generates infinitely many solutions, in contrast to $\mathit{diff}(\lambda x.F, X) \rightarrow^? 0$ (with simplification).

Another alternative is to use a free variable X for the parameter, as for instance done in [SS86]:

$$\begin{array}{rcl} \mathit{diff}''(X,X) & \rightarrow & 1 \\ \mathit{diff}''(\mathit{sin}(X),X) & \rightarrow & \mathit{cos}(X) \\ & \vdots & \end{array}$$

This version is only intended for differentiating ground terms. There is a further problem, as it is unclear how to handle constant functions. A rule $\mathit{diff}''(0,X) \rightarrow 0$ immediately destroys confluence.

8.1.3 A Functional-Logic Parser

Top-down parsers belong to the classical examples for logic programming. The support for non-determinism in logic programming is the main ingredient for this application. On the other hand, functional parsers (see e.g. [Pau91]) have other benefits, such as abstraction over parsing functions. We will integrate the best of both approaches in this example.

We model the following tiny grammar, which is similar to [Pau91]. A grammar is described by the following terms. In addition the constructs $\mathit{and}(T,T')$, $\mathit{or}(T,T')$, and $\mathit{rep}(T)$, there are terminal symbols, e.g. a, b, c, which are recognized via an auxiliary function t. Their meaning is shown in the following table.

Construct	recognizes
t(a)	a, where a is a terminal symbol
$\mathit{and}(T,T')$	wv if T recognizes w and T' recognizes v
$\mathit{or}(T,T')$	v if T or T' recognizes v
$\mathit{rep}(T)$	$v_1 \ldots v_n$ if T recognizes each v_i

For example, $\mathit{and}(t(a), \mathit{or}(t(b), t(c)))$ recognizes $[a,b]$ and $[a,c]$. In our setting, each of these constructs is represented as a parsing function (of the same name). The main issue of this example is to show how to model non-deterministic constructs such as the parsing function for *or* with confluent rewrite rules. The solution is to add an extra argument, called prophecy, to *or*, which determines the choice. When invoking *or* with a free variable as prophecy, the desired effect is archived and, in addition, the prophecies tell us which choice was made at each *or* construct.

In the following rules, T, T' represent parsing expressions and L is the input list of terminals. For uniformity we use prophecies for all parsing constructs, and not only for *or*. The constant symbols $pand, p1, p2, pt$ are used as prophecies. The function $t(x)$ is used to parse the terminal symbol x.

$$
\begin{array}{lcl}
t(X,pt,[Y|L]) & \rightarrow & L \quad \Leftarrow X =_s Y \\
and(T,T',pand(P1,P2),L) & \rightarrow & \text{let } T(P1,L) \text{ in } \lambda l.T'(P2,l) \\
or(T,T',por1(P),L) & \rightarrow & T(P,L) \\
or(T,T',por2(P),L) & \rightarrow & T'(P,L) \\
rep(T,pemp,L) & \rightarrow & L \\
rep(T,prep(P,P1),L) & \rightarrow & \text{let } T(P,L) \text{ in } \lambda l.T'(P2,l)
\end{array}
$$

The following example illustrates how parsing is performed:

$$and(t(a),or(t(b),t(c)),P,[a,b]) \rightarrow^? []$$

succeeds with

$$P \mapsto pand(pt,por1(pt)).$$

This solution for the prophecy can be seen as a parsing script showing how the word was parsed. Here, the first choice in the *or* construct was chosen.

As another example consider the goal

$$rep(or(t(b),t(c)),P,[b,c,b]) \rightarrow^? []$$

whose parsing function accepts words of b's and c's. The goal succeeds here with

$$P \mapsto prep(por1(t),prep(por2(t),prep(por1(t),pemp))).$$

Note that it is possible to write goals where the grammar is looked for or is only partially given.

In purely functional versions of this parser, the mechanism needed for non-determinism and search has to be coded by some means. Compared to first-order logic programming, we achieve a higher level of abstraction and flexibility. For instance, the *rep* construct can be used with any parsing function (of the right type). In a first-order version, the constructs *and*, *or* etc would be represented as a data structure, and a function/predicate has to be written to interpret such constructs. Thus, when passing such a data structure, representing a parser, to some other function, this function has to know how to interpret the structure.

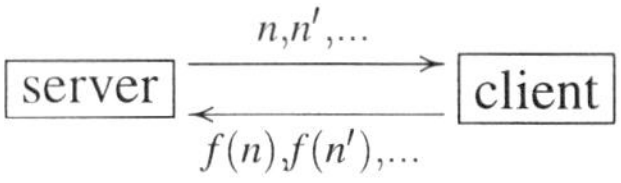

Figure 8.3: Communication Model for Authorization

8.1.4 A Simple Encryption Problem

This example deals with a simple method for authorization. We show that encoding functions can be nicely modeled in our higher-order setting and examine a simple attack on the authorization protocol.

Assume the following method for the authorization of a client and some server along two channels as shown in Figure 8.3. Both parties share an encryption function f. For authorization, a client receives some name a and sends its encryption $f(a)$ to the server. This value $f(a)$ can be viewed as a "password".

For several authorization steps, the channels between the client and the server transmit a list of names and a list of the corresponding passwords. Since the channel is unsafe, the client and the server use the following method to change the password of a name after each use. For simplicity, we assume names are natural numbers. If the client uses the name n and its password $f(n)$, both parties compute a new encryption function f' from f by: $f'(n) = f(n+1)$ and $f'(n+1) = f(n)$. That is, two passwords are swapped.

In the following program, the function $encode(F, List)$ models the client, computing a list of passwords from a list of names of the communications on the channel. It maps the encryption function F to the first element of a stream, and updates the encryption function for the rest of the list.

$$
\begin{array}{lll}
comp(F,G)(X) & \rightarrow & G(F(X)) \\
\\
swap(X,Y,Z) & \rightarrow & if\ Z =_s X\ then\ Y\ else \\
 & & \quad if\ Z =_s Y\ then\ X\ else\ Z \\
\\
encode(F,[X \mid Rest]) & \rightarrow & [F(X) \mid \\
 & & \quad encode(comp(swap(X,X+1),F),Rest)] \\
encode(F,[]) & \rightarrow & [] \\
\\
if\ true\ then\ X\ else\ Y & \rightarrow & X \\
if\ false\ then\ X\ else\ Y & \rightarrow & Y
\end{array}
$$

For instance, if the initial encoding is the identity function, we obtain

$$encode(\lambda x.x, [1,2,2]) = [1,1,3].$$

Now we consider the following situation. Some spy on the channel does not know the initial encryption function, but the method for the update. If the spy observes some communication, his goal is to infer passwords. For instance, assume the spy observes the names $[1,2,2]$ and the corresponding passwords $[a,a,b]$. Then if 3 is sent as the fourth name, we can compute the fourth password with the goal

$$\forall f.encode(f, [1,2,2,3]) \rightarrow^{?} F(f)$$

with solution

$$\theta = \{F \mapsto \lambda f.[f(1), f(1), f(3), f(3)]\}.$$

Clearly $f(1) = a$ and $f(3) = b$ and the spy can infer the fourth password. The encryption in this example is rather simple. It is clearly possible to model more complicated authorization strategies with this approach.

8.1.5 "Infinite" Data-Structures and Eager Evaluation

Infinite data structures are one of the nice features of lazy functional programming, e.g. [Tur86, PHA+96]. We show in the following that such infinite structures can be modeled with terminating functional-logic programs. Although some functional-logic languages, for instance [MNRA92], support non-terminating rules to allow infinite data structures, we argue here that this is not needed in our narrowing context. Recall that termination is essential for many refinements of the narrowing calculi. Also, a drawback of non-terminating rule is that lazy evaluation is strictly needed to avoid possible divergence. Hence this disallows optimizations like normalization.

Consider the example of an infinite list of ones, defined by:

$$ones \rightarrow [1|ones]$$

This rule can be used with rules such as:

$$\begin{array}{rcl} first([X|R]) & \rightarrow & X \\ rest([X|R]) & \rightarrow & R \\ sum\_n(0,L) & \rightarrow & 0 \\ sum\_n(s(n),[X|R]) & \rightarrow & X + sum\_n(n,R) \end{array}$$

Lazy evaluation, which only evaluates subterms when needed, yields for instance

$$sum\_n(4, ones) \longrightarrow 4$$

This model of lazy computation has the disadvantage that non-terminating rules, here $ones \to [1\,|ones]$, have to be used carefully to avoid divergence.

We can model such infinite structures with terminating rules in our setting. We simply reverse the rule generating infinite objects:

$$ones([1|R]) \quad \to \quad [1|ones(R)]$$

The technique for working with this definition is as follows: given an object of appropriate, but unknown size, compute the solution. Thus, terminating rules suffice and eager reduction (or normalization in rewriting parlance) is possible.

Using the above definition, we can state the query

$$sum\_n(4, ones(Y)) \to^? 4,$$

which has the desired solution

$$\{Y \mapsto [1,1,1,1|Y'], \ldots\}.$$

Thus the term $ones(Y)$ represents an "infinite" list.

Call-by-need narrowing is particularly useful in this example, as it solves goals only when needed. In the above example, intermediate goals of the form $ones(Y) \to X$ are simply delayed. Only if X is instantiated, the goal is simplified and possibly delayed again.

The above example only models functional programming, which aims at evaluating expressions to unique values. Compared to functional programming, this approach also models search as in logic programming. For instance, lists where each element is a one or a two are easy to model. This is not possible with the functional approach, as it would require non-confluent rules. Note that this technique for modeling infinite functional structures appears in several of the examples, for instance in the hardware synthesis example.

8.1.6 Functional Difference Lists

Difference lists are a standard technique [SS86] for implementing lists in logic programming such that appending two lists can be done in constant time. A difference list is a pair, where the first element is the actual list of interest and the second element is the tail of the first list, typically a free variable. For instance, to represent the list $[a,b,c]$ as a difference list, we use the pair $([a,b,c \mid R],R)$ for some variable R.

Concatenating two difference lists is done in functional-logic programming by the function

$$append((X,Y),(Y,R)) = (X,R)$$

and in plain logic programming by the corresponding predicate. A principal problem with this representation is that a concrete variable is used to represent the end of the list. Thus when copying a difference list, a "predicate" is needed to introduce a new variable at the end of the copied list, as e.g. shown in [Red94]. This is not only rather involved but also takes linear time.

The functional equivalent is to abstract over this variable representing the rest of the list. Thus we use functions from lists to lists as "functional difference lists", i.e. $\lambda x.[a,b,c \mid x]$ instead of $[a,b,c \mid X]$. This idea was introduced by Hughes [Hug86] and compared to the logic approach by Burton [Bur89] and Reddy [Red94]. We believe that this representation is much clearer than using free variables, which must be "new" for each copy. For instance, appending two functions is straightforward by β-reduction:

$$append1(L,R) \rightarrow \lambda x.L(R(x))$$

A nice result on this approach in higher-order logic programming was shown in [BR91]: a naive reverse function on lists can be linear.

It appears that this implementation of lists is also possible in functional languages. However, since these do not permit rewrite rules with abstractions on the left, simple operations become tedious. For instance, computing the first element of a list represented as $\lambda x.[a,b,c \mid x]$, is done here by a rewrite rule

$$first(\lambda x.[X, Rest(x)]) \rightarrow X.$$

In a usual functional language, one cannot access terms below abstractions directly and an extra application is needed beforehand.

8.1.7 The Alternating Bit Protocol

This example shows how to model distributed systems with inherent non-determinism by functional-logic programming. For non-deterministic algebraic specifications, this has been shown in [Huß93]. We model the inherent non-determinism of distributed systems via logic variables. The alternating bit protocol [BSW69] is used for reliable communication on unreliable asynchronous channels. Assume sender s and receiver r are connected as in

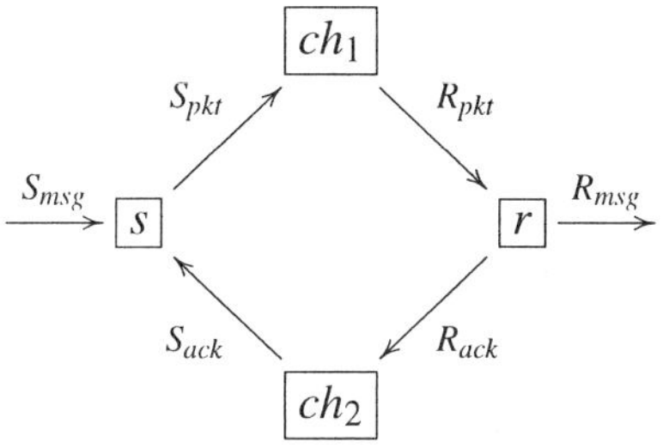

Figure 8.4: Alternating Bit Protocol

Figure 8.4 via the transmitters ch_1 and ch_2. These model unreliable communication, as they may lose messages. The goal is to hide this from the environment and to deliver exactly the messages on R_{msg} which have been received via S_{msg}.

For safe communication, s sends incoming packages together with a marking bit b. If r receives a message, only the attached bit is sent back via ch_2. Sender s repeats sending a message until the same bit is received at S_{ack}. Then s starts sending the next message marked with $\neg b$. Similarly, r only sends a message on R_{msg}, if the bit changes.

We will model the communication histories via streams, one for each channel. Note that this model implicitly provides for asynchronous communications via buffering messages for each channel. Each component is modeled as a function from input to output streams. There is however a further complication with this model, as discussed in [Bro93]: we need to model time progress explicitly. Roughly speaking, the sender must wait for answers but also repeat sending messages. However, this is not possible without some notion of time progress. More formally, computable functions which are monotonic w.r.t. the prefix order on streams do not suffice here [Bro93]. Hence we introduce an extra message $\surd$ which models the advance of time, also called time tick. The transmitters ch_1 and ch_2 are modeled as functions which take an additional parameter, the prophecy. These streams determine the loss of data and may also send time ticks. More concretely, the channels read a data object from the input and pass it on if the boolean value on the prophecy channel is true.

For the following function definitions in Figure 8.5, the relation between argument positions and the channel names in Figure 8.4 can be seen in the goals below. Only for ch_1 and ch_2, the second arguments are prophecies, which are not shown in Figure 8.4. As the receiver has two output channels,

$$
\begin{array}{lcl}
s(U,V) & \to & s'(true,U,V) \\
s'(B,[],V) & \to & [] \\
s'(B,[X|U],[\surd|V]) & \to & [(X,B)|s'(B,[X|U],V)] \\
s'(B,[X|U],[B'|V]) & \to & [(X,B)|s'(B,[X|U],V)] \quad \Leftarrow B =_s \neg B' \\
s'(B,[X|U],[B'|V]) & \to & s'(\neg B,U,V) \quad \Leftarrow B =_s B' \\
r(U) & \to & r'(false,U) \\
r'(B,[]) & \to & ([],[]) \\
r'(B,[\surd|U]) & \to & let\ (U',V') = r'(B,U)\ in\ ([\surd|U'],[\surd|V']) \\
r'(B,[(X,B')|U]) & \to & let\ (U',V') = r'(\neg B,U)\ in\ ([X|U'],[B|V']) \\
 & & \Leftarrow B =_s B' \\
r'(B,[(X,B')|U]) & \to & let\ (U',V') = r'(B,U)\ in\ (U',[\neg B|V']) \\
 & & \Leftarrow B =_s \neg B' \\
ch_1(S,P) & \to & ch(S,P) \\
ch_2(S,P) & \to & ch(S,P) \\
ch([],P) & \to & [] \\
ch([X|U],[\surd|P]) & \to & [\surd|ch([X|U],P)] \\
ch([X|U],[true|P]) & \to & [X|ch(U,P)] \\
ch([X|U],[false|P]) & \to & [\surd|ch(U,P)]
\end{array}
$$

Figure 8.5: Rules for the Alternating Bit Protocol

we use a pair of streams for this. Note that sender and receiver must remember a marking bit, for which both are embedded into functions with an extra parameter.

With this specification, we can model the system via the following goal list:

$$
\begin{array}{c}
s(S_{msg},S_{ack}) \to^? S_{pkt}, \\
ch_1(S_{pkt},P_1) \to^? R_{pkt}, \\
r(R_{pkt}) \to^? (R_{msg},R_{ack}), \\
ch_2(R_{ack},P_2) \to^? S_{ack}
\end{array}
$$

For simulation, we need to instantiate some of the variables. As an example, consider the following input

$$S_{msg} = [d_1,d_2], P_2 = [true,true], P_2 = [\surd,true,true].$$

Note that sending a time tick on S_{ack} (via P_2) is used to start the transmission.

This produces the following solution (or behavior):

$$\begin{aligned}&\{S_{pkt} \mapsto [(d_1, true), (d_2, false)],\\ &R_{pkt} \mapsto S_{pkt},\\ &R_{msg} \mapsto [d_1, d_2],\\ &S_{ack} \mapsto [\surd, true, false],\\ &R_{ack} \mapsto [true, false]\}.\end{aligned}$$

To model some loss of data, we can try

$$\begin{aligned}&\{S_{msg} = [d_1|d_2],\\ &P_1 = [false|true|true],\\ &P_2 = [\surd|true|true|true]\}\end{aligned}$$

which yields

$$\begin{aligned}&\{S_{pkt} \mapsto [(d_1, true), (d_1, true), (d_2, false)],\\ &R_{pkt} \mapsto [\surd, (d_1, true), (d_2, false)],\\ &R_{msg} \mapsto [d_1, d_2],\\ &S_{ack} \mapsto [\surd, \surd, true, false],\\ &R_{ack} \mapsto [\surd, true, false]\}.\end{aligned}$$

Clearly, many more interesting simulations are possible. In general, only correct outputs on R_{msg} should be computed for all solutions for P_1 and P_2. Note that functional-logic programming is not powerful enough for formal verification, but is complete w.r.t. finding counter examples, which can be very useful during development. Full verification of the algorithm (see [Bro93]) is of course more involved and is usually attempted after simulation.

8.2 Equational Reasoning by Narrowing

This section presents two examples where full equational reasoning is needed. More precisely, we need full equality, which cannot be encoded with left-linear rules. This, for instance, is useful to reason about functional programs, as shown in the next example.

8.2.1 Program Transformation

The utility of higher-order unification for program transformations has been shown nicely by Huet and Lang [HL78] and has been developed further in [PE88, HM88]. This example for unfold/fold program transformation is taken from [FH88]. It is in fact an instance of deforestation [Wad90]. We assume the following standard rules for lists

$$\begin{aligned} map(F,[X|R]) &\rightarrow [F(X)|map(F,R)] \\ foldl(G,[X|R]) &\rightarrow G(X, foldl(G,R)) \end{aligned}$$

Now assume writing a function $g(F,L)$ by

$$g(F,L) \rightarrow foldl(\lambda x,y.plus(x,y), map(F,L))$$

that first maps F onto a list and then adds the elements via the function *plus*. This simple implementation for g is inefficient, since the list must be traversed twice. The goal is now to find an equivalent function definition that is more efficient. We can specify the recursive case for g with higher-order terms in a syntactic fashion by one simple equation:

$$\forall f,x,l.g(f,[x|l]) =^{?} B(f(x), g(f,l))$$

The variable B represents the body of the function to be computed and the first argument of B allows the use of $f(x)$ in the body. The scheme on the right only allows recursing on l for g.

To solve this equation, we add a rule $X =^{?} X \rightarrow true$ as described in Section 6.5.1, and then apply narrowing, which yields the solution $\theta = \{B \mapsto \lambda fx, rec.plus(fx, rec)\}$ where

$$g(f,[x|l]) = \theta B(f(x), g(f,l)) = plus(f(x), g(f,l)).$$

This shows the more efficient definition of g. In this example, simplification can reduce the search space for narrowing drastically: it suffices to simplify the goal to

$$\begin{aligned} \lambda f,x,l.plus(f(x), foldl(plus, map(f,l))) &=^{?} \\ \lambda f,x,l.B(f(x), foldl(plus, map(f,l))), & \end{aligned}$$

where narrowing with the newly added rule $X =^{?} X \rightarrow true$ yields the two goals

$$\begin{aligned} \lambda f,x,l.plus(f(x), foldl(plus, map(f,l))) &\rightarrow^{?} \lambda f,x,l.X(f,x,l), \\ \lambda f,x,l.B(f(x), foldl(plus, map(f,l))) &\rightarrow^{?} \lambda f,x,l.X(f,x,l). \end{aligned}$$

These can be solved by pure higher-order unification. It should be noted that our notion of oriented goals requires an additional rule for equality. In an implementation it is desirable to hide such details from the user.

8.2.2 Higher-Order Abstract Syntax: Type Inference

In this section, we consider the problem of polymorphic type reconstruction as it occurs in functional languages, see for instance [CDDK86, NP95]. For simplicity, we only consider the core constructs of such a language, i.e. typed λ-calculus. The syntax of the language includes atoms $const(x)$, application $app(t,t')$, and abstraction $abs(\lambda x.t)$.

The set of polymorphic types is generated by some base types, type variables, and the function type constructor $->$, written in infix notation as $\sigma -> \tau$. We chose the symbol $->$ instead of the common $\rightarrow$ in order to avoid confusion with term rewriting or with types in the underlying logic.

For instance, a term $succ(5)$, where $succ$ is a function on integers, is represented as

$$app(const(succ), const(5)).$$

As usual for type inference systems, we store the type of atoms in a context E. For instance, compared to the typing rules of simply typed λ-calculus in Section 3.2, the judgment $x:\tau$ in the rule

$$\frac{x:\tau \quad s:\tau'}{(\lambda x.t):(\tau -> \tau')}$$

is represented in a context.

In violation of our conventions, we write free variables over types by Greek letters σ and τ as usual for type inference systems. The standard rules for type inference can easily be expressed as conditional equations:

$$\begin{array}{lll}
update(E,T,X,Y) & \rightarrow & if\ Y =^? X\ then\ T\ else\ E(Y) \\
\\
type_rel(E,const(X),E(X)) & \rightarrow & true \\
type_rel(E,app(T,T'),type(\tau)) & \rightarrow & true \\
 & \Leftarrow & type_rel(E,T,type(\sigma -> \tau)) \rightarrow true, \\
 & & type_rel(E,T',type(\sigma)) \rightarrow true \\
type_rel(E,abs(\lambda x.T(x)), & & \\
\quad type(\sigma -> \tau) & \rightarrow & true \\
 & \Leftarrow & \lambda v.type_rel(update(E,type(\sigma),v), \\
 & & \qquad T(const(v)),type(\tau)) \rightarrow \lambda v.true
\end{array}$$

The function $update(E,T,X)$ creates a new context where X has the type T. Notice that the last two rules have the extra free variables σ and τ that do not occur on the left-hand side. In the last rule for typing an abstraction, characteristic for higher-order abstract syntax, a local constant v serves for exploring the type of $\lambda x.T(x)$.

For example, a term $f(\lambda x.plus(x,y))$, where f and y are polymorphic atoms, is represented as

$$t = app(const(f), abs(\lambda x.app(app(const(plus), x), const(y)))).$$

Type inference for t is done by the following query

$$type\_rel(E, t, type(\beta)) \to^{?} true,$$

where $E(f) = \alpha \mathrel{-\!\!>} \alpha$, $E(y) = \gamma$, and $E(plus) = int \mathrel{-\!\!>} int \mathrel{-\!\!>} int$. This goal has the solution

$$\{\beta \mapsto int \mathrel{-\!\!>} int, \alpha \mapsto int \mathrel{-\!\!>} int, \gamma \mapsto int, \ldots\}.$$

Using higher-order terms to describe programs or formal systems has also been called higher-order abstract syntax. For more examples in this direction, we refer to [PE88, MP92b].

In the above rules, the extra variables are purposely introduced to compute "local" types. For instance, the variable σ in the abstraction rule only serves for computing the type of the subterm T. Thus rewriting requires computing solutions for these variables in the conditions, as done by narrowing. Notice that Simple Systems do not suffice for this example, as the rules are not left-linear. This example requires full unification, as the occurs check is needed for safe type inference. Hence it seems impossible to encode this example easily by left-linear rules.

Chapter 9

Concluding Remarks

This work was motivated by the idea that higher-order equations can be used in practical systems for equational reasoning and functional-logic programming. Towards this goal we have first examined decidable classes of higher-order unification. We have shown that for many practical purposes, higher-order unification is not only a powerful tool, but also terminates for several classes of terms. The main restriction we need is linearity, which is common in programming. It also explains to some degree why higher-order unification in logic programming [NM88] and higher-order theorem proving [Pau94, AINP90] rarely diverges. Secondly, we have developed a framework for solving higher-order equations by narrowing. For lazy narrowing, we were able to develop many important refinements, such as normalization and eager variable elimination for normalized solutions. Of similar practical importance are the extensions to conditional equations. We have also seen that some approaches such as plain narrowing are not suitable for the higher-order case.

The work on left-linear rewrite systems for programming applications led to Simple Systems, which is an important class of goals for equational programming. This class enjoys many useful properties, for instance solved forms are easy to detect. Furthermore, in the second-order case, unification remains decidable for Simple Systems. Simple Systems are a large class of goals where the occurs check is not needed. Interestingly, in most implementations of Prolog, the occurs check is missing for computational reasons. This suggests the definition of languages where the occurs check is redundant. More importantly, it indicates that the problems solvable with Prolog implementations correspond to such a class of problems. The main result for Simple Systems is the strategy of call-by-need narrowing, where intermediate goals can be identified and can safely be delayed, which leads to an effective computation strategy.

Altogether, we believe that the results for normalized solutions and Simple Systems are a major step towards high-level programming languages, where efficiency and a simple operational model are significant. This leads to a novel approach to functional-logic programming that is oriented towards higher-order functional languages. Whereas some approaches aim at extending logic programming by functions, the main idea here is to extend a higher-order functional language by logic or free variables as in Prolog. This approach facilitates several operational optimizations which are not possible in other approaches oriented towards extending logic programming.

Another observation is that oriented goals turn out to be a particularly useful restriction on equational goals. Oriented goals do not limit the expressiveness, i.e. full unification can be encoded, and simplify the technical treatment. Furthermore, we have seen that for left-linear HRS, there is a difference between matching, as performed by oriented goals, and full unification. In the former, Simple Systems suffice for narrowing and the decidability of second-order unification is maintained.

This work also contributes to equational reasoning in higher-order theorem provers. For this, we have provided complete calculi for higher-order narrowing for convergent rewrite rules. Although for most of the optimizations we considered, some restrictions were needed, they apply to general theorem proving. As another application, Section 5 can be the basis for further investigation into the decidability of second-order R-matching problems. For instance, Curien [Cur93] presents first results on second-order E-matching for first-order E.

9.1 Related Work

Since this work aims at integrating and generalizing several research directions, we compare it to similar approaches and other paradigms in the following. The classification into narrowing and functional-logic programming is somewhat artificial, as many of the quoted languages and technical results fall into both.

9.1.1 First-Order Narrowing

For the theory of first-order narrowing, it took about fifteen years and many contributions to develop optimal strategies, see for instance [Han94b] for an overview. The first of these, presented in [BKW93], assures that no solution is computed twice for convergent systems, but requires considerable runtime overhead. For a restricted class of rewrite rules, i.e. inductively sequential, an optimal narrowing strategy, called needed narrowing, has been presented in [AEH94].

Needed narrowing in [AEH94] is optimal w.r.t. the length of the reduction steps performed (modulo sharing; for a precise definition see [HL91]) and does not produce a solution twice. Only the latter optimality result has been obtained in [HP96] for restricted higher-order rewrite rules (constructor-based and inductively sequential). In the higher-order case, optimal reduction strategies are a topic of current research [Oos96]. Hence we only model needed reduction here, which may not be optimal, but is used in existing functional languages.

Note that optimality of reductions is not the only criterion for comparing narrowing strategies. One advantage in our setting is that the right-hand sides of goals can be used to discover deterministic operations or failure cases. For instance, with the following rules

$$\begin{array}{lll} f(0) & \rightarrow & 0 \\ ones(0) & \rightarrow & s(0) \\ ones(s(X)) & \rightarrow & ones(X) \end{array}$$

the goal $f(ones(X)) \rightarrow^? s(0)$ obviously has no solution. This is detected here by a constructor clash during simplification (see Sec. 6.3.3). On the other hand, a strategy driven purely by optimal reductions, such as in [AEH94], attempts narrowing steps at the inner $ones(X)$ subterm and diverges. Although it is difficult to compare these approaches for practical applications, notice that Prolog, when coding the above into predicates, performs the same simplification, i.e. unification fails.

A disadvantage of the outside-in approach of lazy narrowing is that redundant computations in different search trees are possible. For instance, consider the goal $ones(t) \rightarrow^? s(Y)$, where t is an arbitrary term. In this example, for each of the two rules, the term t has to be evaluated, i.e. in two goals $t \rightarrow^? 0$ and $t \rightarrow^? s(X)$. This can be avoided by needed narrowing [AEH94]. Again, a naive translation into Prolog may exhibit the same inefficiency, depending on the order of the literals. Also, this may not be a problem for mostly functional programs with little branching.

9.1.2 Other Work on Higher-Order Narrowing

There have been several earlier works on higher-order narrowing, but none covering the full higher-order case. Qian [Qia94] lifted the completeness of first-order narrowing strategies to higher-order patterns for first-order rules. Higher-order patterns are an important subclass of λ-terms, which include bound variables and behave almost as first-order terms in most respects. However, patterns are often too restrictive, as obvious from the examples in Sections 2.8 and Chapter 8 (see also [Pre94b, MP92a]). In particular, they do not suffice for modeling higher-order functional programs. Examples are the function *map* in Section 2.7 and the definition of the let-construct by a higher-order rewrite rule

$$\mathsf{let}\ X\ \mathsf{in}\ T \to T(X)$$

where the right-hand side is not a pattern. Thus rewriting or narrowing with this rule may introduce non-pattern terms. For a discussion on this issue in a (higher-order) logic-programming context see [MP93].

The approach to higher-order narrowing in [LS93, ALS94a] aims at narrowing with higher-order functional programs and does not limit rules to higher-order patterns. Rules with pattern left-hand sides are used for narrowing on quasi-first-order terms. (These are slightly more general than quasi first-order terms defined here.) Although this seems to be an interesting compromise, it has strong restrictions: higher-order variables in the left-hand sides of rules may occur only directly below the outermost symbol. This guarantees that the resulting term is still quasi-first-order. For instance, the function $map(F,[X|R])) \to \ldots$, fulfills this requirement only if X and Y are first-order. The following slight extension of the map example,

$$map2(pair(F_1,F_2),[pair(X,Y)|R])) \;\to\; [pair(F_1(X),F_2(Y))| \\ map2(pair(F_1,F_2),R)]$$

is for instance not permitted.

Roughly speaking, when narrowing with such rules, narrowing and rewriting coincide for these higher-order variables as they occur only at depth one on the left-hand side.[1] This restriction to avoid full higher-order narrowing is also used in a similar form in [GHR92].

Other recent works assume applicative rewriting to model higher-order constructs, which can be interpreted as modeling function application by a function *apply*. This yields a simple first-order model of higher-order functions (without abstractions). For instance, we can write the map function

$$map(F,[X|R]) \to [F(X)|map(F,R)]$$

in terms of a binary application operator *apply* for concrete function symbols:

$$apply(apply(map,F),[X|R]) \to [apply(F,X)|apply(apply(map,F),R)]$$

[1] This result is a bit more subtle, since the rewrite rules are lifted for narrowing. Lifting turns a first-order term into a higher-order term. This problem is however not addressed in [LS93, ALS94a].

Note that in this way higher-order function symbols are treated as first-order constants.

The main drawback of this approach is that functional variables denote just one concrete function symbol or an *apply* term. Abstractions are not possible in solutions and thus extensionality (via η-conversion) is lost. This approach is pursued in an untyped, first-order like setting in [NMI95] and in [GHR97] for a particular design of a functional-logic language. This simple approach to higher-order functions has been advocated for logic programming [War82]. Also, many Prolog implementations support a quasi second-order predicate "apply" which applies a variable function symbol to some arguments. Use of this built-in predicate obviously destroys completeness in the usual sense.

9.1.3 Functional-Logic Programming

All of the many higher-order extensions of functional-logic languages [BG86, She90, CKW89, GHR92, Loc93, Llo94, AKLN87, GHR97, NMI95] are, to our knowledge, limited to first-order unification and are not complete in a higher-order sense. For instance, the work in [Loc93] uses higher-order variables, but only (first-order) narrowing on first-order terms plus β-reduction as the operational model. Since higher-order rules such as *map* are used, higher-order terms can be created which cannot be handled by this approach. The work in [GHR92] on SFL, an extension of the language BABEL [MNRA92], similarly permits higher-order variables. Completeness of narrowing with first-order unification is claimed w.r.t. particular denotational and operational semantics for partial objects.

Other currently developed functional-logic languages such as Oz [NS94, Smo94], Escher [Llo94] and Curry [HKMN95] do not utilize higher-order unification and hence do not focus on completeness as in logic programming. For instance, Oz focuses on concurrent and constraint programming and provides no build-in search mechanisms. Since the logic programming part of Oz is first-order, it treats functional objects as closures and maintains references to them. These can be manipulated as first-order data-types, but have the drawback that equality of functions is just pointer equality. Whereas Oz employs an abundance of language extensions, the language is Escher fully declarative, based on higher-order logic. Its computational mechanism is, roughly speaking, higher-order rewriting. A formal computation model for functional logic programming with suspensions as in current logic and constraint programming has been proposed in [Han97]. It will be the underlying computation model for the currently developed language Curry [HKMN95].

9.1.4 Functional Programming

It is instructive to elaborate the relation of higher-order rewriting and functional programming languages. First of all, functional languages do not permit λ-terms as data structures. But even for plain functional programs, there is a difference. The point is that the latter do not implement some typed λ-calculus with recursion via the Y-combinator, but implement a weak λ-calculus [Plo75, Mar90]. The idea behind weak λ-calculi is to disallow reductions below λ-abstractions. This suffices for programming, since only first-order terms can be explicitly computed. Interestingly, weak λ-calculi are not confluent. The reason for avoiding reductions below binders is clearly efficiency. In our context, it is often essential to compute below binders. For instance, it is clearly necessary for solving $\forall$-quantified goals and for computing/reasoning with functional objects.

9.1.5 Higher-Order Logic Programming

Higher-order logic programming [NM88] employs two major extensions of first-order logic programming: first, higher-order terms are used and, secondly, hereditary Harrop formulas, which generalize horn clauses.

The latter, roughly speaking, allow for local implication and universal quantification in goals and in the premises of clauses. This can be modeled here by conditions, since it is is possible to have conditions with local quantification via λ-abstractions such as

$$l \rightarrow r \Leftarrow \forall x.p(x) \rightarrow true$$

The other new notion, of local implication allows the modeling of "local" rules. For instance, to prove

$$(Q \Rightarrow P)$$

P must be proven with Q as an additional assumption. This nice extension is not directly entangled with the higher-order aspects. It could possibly be modeled by some extensions of (conditional) narrowing.

Compared to higher-order logic programming, the functional approach with Simple Systems lies between λ-Prolog [NM88], where full higher-order terms are used, and Elf [Pfe91],[2] where non-patterns are just delayed as constraints. The main advantage of this functional approach is that a decidable class of second-order unification instead of pattern unification can be used. The problem is that higher-order patterns cover large classes of terms occurring in practice, but are not sufficient in general. Thus strategies have been developed in [MP93] to handle such cases effectively.

[2]It should be noted that Elf has a much more expressive type system.

Another difference to (higher-order) logic programming is that predicates and terms are not separated. Higher-order λ-terms are used for data structures and do not permit higher-order programming as in functional languages. For instance, the function *map* in Section 2.7 cannot be written directly by higher-order logic programming. Nadathur [Nad87] reports similar problems with variables over predicates for modeling *map* in higher-order logic programming. In this respect, our approach is more general and allows for an arbitrary integration of data and functions.

9.2 Further Work

The next crucial step for our new framework is to develop efficient implementations, or to find interesting fragments which can be implemented efficiently. Also, several other extensions, as developed in related contexts, are of interest.

9.2.1 Implementation Issues

Developing efficient implementations for higher-order programming is a challenging task. First steps in this direction are an abstract machine for higher-order logic programming [NJW93] and a compiler for higher-order logic programming [BR92].

In general, the difficult problem is to implement λ-terms. There exist several approaches for representing the concepts of λ-calculus. The first idea for representing binders and bound variables goes back to de Bruijn [dB72]. Bound variables are represented as natural numbers, where the number indicates the corresponding binder: the de Bruijn index i refers to the i-th abstraction when going up in the term structure. In other words, it counts the number of abstractions between the occurrence and its binder. This for instance entails that two different occurrences of a bound variable may be translated to different numbers. For the usual conversions and operations, such as substitutions, the binders have to be adapted, which can be complex and time consuming. A more general framework for such problems are calculi with explicit substitutions, e.g. [ACCL91, BBLRD96], which support abstract operations for adapting binders. Some of these ideas are for instance used by Nadathur for implementing λProlog [NJW93]. For higher-order pattern unification, such representations are developed in [Nip93a] and for full higher-order unification in [DHK95]. The latter work for instance attempts to optimize the index computations needed for substitutions. One of the advantages of such a representation of λ-terms is that some operations, such as lifting, can be implemented efficiently. For instance, with de Bruijn indices, lifting can be performed by integer comparison in implementations of λProlog [Mil91a].

As an alternative to de Bruijn indices, de Bruijn levels can be used. The idea is that some index i representing a bound variable refers to the i-th abstraction occurring on the path to its occurrence. Then, in particular, all occurrences of the same bound variable enjoy the same index. The drawback is that when substituting a term t for some variable in some other term, the indices of local variables in t have to be adapted. A framework for dealing with this approach is discussed in [LRD95].

Whereas the above methods for manipulating higher-order terms as first-order data structures still reflect the terms structure of the corresponding λ-term, it is possible to encode higher-order terms directly into first-order terms without a notion of bound variables. This alternative, called combinatory logic [HS86], models the reduction of λ-calculus by some particular first-order rewrite rules. Combinatory logic is a common choice for implementing functional languages, see for instance [Jon92], where reduction is of interest. For more general equational reasoning, there also exist first attempts [DM93, DJ95]. One of the problems of this approach is that the notion of higher-order patterns and the decidability results for higher-order unification have not been transferred to this setting. An implementation for a higher-order functional logic language based on combinators was proposed in [KA96], but a detailed comparison to related work is missing.

9.2.2 Other Extensions

Several extensions which have been developed for equational reasoning and functional-logic languages are of interest in our context. Some of these are discussed in the following.

More expressive type systems, such as polymorphism and type classes in current functional languages [NP95], have not been considered here. An extension to polymorphism faces the problem that higher-order unification with polymorphism is infinitely branching [Nip91b]. This stems from the fact that the imitation rule of higher-order unification depends on the typing. Hence, if the type is unknown as it is polymorphic, the simple solution is infinite branching. In practice, such cases are rare as experience with λProlog and with the Isabelle system [Pau94] shows. An attempt to solve this problem using calculi with explicit substitutions is currently explored [DHK95, BBLRD96].

Several further issues have not been examined here. For instance, the restriction to terminating rules is hard to enforce in a practical language. A first step towards optimality was presented in [HP96] for a more restricted setting, where optimality w.r.t. solutions is achieved. This means that only incomparable solutions are computed.

As we have focused on equational logic, we did not cover inductive theorems, which are pervasive in theorem proving applications. Inductive equations only hold for initial structures, either defined explicitly for inductive data structures or implicitly via initial models. In the first-order case, there is an elegant link between both approaches, which allows effective methods for inductive proofs in the presence of convergent rewrite rules. This is known as induction-less induction [Bac91] and employs techniques similar to narrowing. Although model theory is more involved in the higher-order case [And86], there is a first, more syntactic approach for higher-order induction-less induction in [LPL96].

Another interesting application is to model calculi for distributed systems, e.g. the π-calculus [MPW92a, MPW92b], by higher-order rewriting, thus obtaining an executable version. The major obstacle for this approach is that most of the rules in this calculus apply modulo associativity (A) and commutativity (C). Rewriting modulo AC has been extensively studied in the first-order case, for the higher-order case there exist first results on AC-unification [QW94, MW94].

Bibliography

[ACCL91] M. Abadi, L. Cardelli, P.-L. Curien, and J.-J. Lévy. Explicit substitutions. *Journal of Functional Programming*, 1(4):375–416, 1991.

[AEH94] S. Antoy, R. Echahed, and M. Hanus. A needed narrowing strategy. In *Proc. 21st ACM Symposium on Principles of Programming Languages*, pages 268–279, Portland, 1994. ACM Press.

[AFM+95] Z. Ariola, M. Felleisen, J. Maraist, M. Odersky, and P. Wadler. A call-by-need lambda calculus. In *22'nd ACM Symposium on Principles of Programming Languages*, San Francisco, California, 1995. ACM Press.

[AINP90] P. Andrews, S. Issar, D. Nesmith, and F. Pfenning. The TPS theorem proving system. In M.E. Stickel, editor, *Proc. 10th Int. Conf. Automated Deduction*, pages 641–642. LNCS 449, 1990.

[AKLN87] H. Aït-Kaci, P. Lincoln, and R. Nasr. Le fun: Logic, equations and functions. In *Proceedings of the Fifth International Conference and Symposium on Logic Programming*, pages 17–23, San Francisco, August - September 1987. IEEE, Computer Society Press.

[ALS94a] J. Avenhaus and C. A. Loría-Sáenz. Higher-order conditional rewriting and narrowing. In J.-P. Jouannaud, editor, *1st International Conference on Constraints in Computational Logics*, München, Germany, September 1994. Springer LNCS 845.

[ALS94b] J. Avenhaus and C. A. Loría-Sáenz. On conditional rewrite systems with extra variables and deterministic logic programs. In *LPAR '94*, Lecture Notes in Computer Science, vol. 822, Kiev, Ukraine, July 1994. Springer-Verlag.

[And86] P. Andrews. *An Introduction to Mathematical Logic and Type Theory: to Truth through Proof.* Computer Science and Applied Mathematics. Academic Press, 1986.

[Bac91] L. Bachmair. *Canonical Equational Proofs.* Progress in Theoretical Computer Science. Birkhäuser, 1991.

[Bar84] H. Barendregt. *The Lambda Calculus, its Syntax and Semantics.* North Holland, 2nd edition, 1984.

[Bax76] L. D. Baxter. *The Complexity of Unification.* PhD thesis, University of Waterloo, Waterloo, Canada, 1976.

[BBLRD96] Z. Benaissa, D. Briaud, P. Lescanne, and J. Rouyer-Degli. $\lambda\nu$, a calculus of explicit substitutions which preserves strong normalisation. *Journal of Functional Programming*, 6(5):699–722, September 1996.

[BG86] P. G. Bosco and E. Giovannetti. IDEAL: An ideal deductive applicative language. In G. Lindstrom and R. M. Keller, editors, *Symposium On Logic Programming*, pages 89–96. IEEE, 1986.

[BG89] H. Bertling and H. Ganzinger. Completion-time optimization of rewrite-time goal solving. In N. Dershowitz, editor, *Proceedings of the Third International Conference on Rewriting Techniques and Applications*, pages 45–58. Springer LNCS 355, 1989.

[BGM88] P. G. Bosco, E. Giovanetti, and C. Moiso. Narrowing vs. SLD-resolution. *Theoretical Computer Science*, 59:3–23, 1988.

[BKW93] A. Bockmayr, S. Krischer, and A. Werner. An optimal narrowing strategy for general canonical systems. In M. Rusinowitch and J. L. Remy, editors, *Conditional Term Rewriting Systems: Proc. of the Third International Workshop (CTRS-92)*, pages 483–497. Springer LNCS 850, 1993.

[BN] F. Baader and T. Nipkow. *Term Rewriting and All That.* Cambridge University Press. To appear.

[BR91] P. Brisset and O. Ridoux. Naïve reverse can be linear. In K. Furukawa, editor, *Proceedings of the Eighth International Conference on Logic Programming*, pages 857–870, Paris, France, 1991. The MIT Press.

[BR92] P. Brisset and O. Ridoux. The architecture of an implementation of lambda-prolog: Prolog/mali. In *Proc. Workshop on LambdaProlog, Philadelphia*, 1992. PA, USA.

[Bra75] D. Brand. Proving theorems with the modification method. *SIAM Journal of Computing*, 4:412–430, 1975.

[Bro93] M. Broy. Functional specification of time sensitive communicating systems. *ACM Transactions on Software Engineering and Methodology*, 2(1):1–46, January 1993.

[BS94] F. Baader and J. Siekmann. Unification theory. In D.M. Gabbay, C.J. Hogger, and J.A. Robinson, editors, *Handbook of Logic in Artificial Intelligence and Logic Programming*. Oxford University Press, 1994.

[BSW69] K. A. Bartlett, R. A. Scantlebury, and P. T. Wilkinson. A note on reliable full-duplex transmission over half-duplex lines. *Communications of the ACM*, 12(5):260–261, 1969.

[BT87] J. A. Bergstra and J. V. Tucker. Algebraic specifications of computable and semicomputable data types. *Theoretical Computer Science*, 50:137–181, 1987.

[Bur89] F. W. Burton. A note on higher-order functions versus logical variables. *Information Processing Letters*, 31:91–95, 1989.

[BvEG$^+$87] H. P. Barendregt, M. C. J. D. van Eekelen, J. R. W. Glauert, J. R. Kennaway, M. J. Plasmeijer, and M. R. Sleep. Term graph rewriting. In J. W. de Bakker, A. J. Nijman, and P. C. Treleaven, editors, *PARLE: Parallel Architectures and Languages Europe (Volume 2: Parallel Languages)*, pages 141–158. Springer LNCS 259., 1987.

[CAB$^+$86] R. Constable, S. Allen, H. Bromly, W. Cleaveland, J. Cremer, R. Harper, D. Howe, T. Knoblock, N. Mendler, P. Panangaden, J. Sasaki, and S. Smith. *Implementing Mathematics With the Nuprl Proof Development System.* Prentice-Hall, New Jersey, 1986.

[CDDK86] D. Clément, J. Despeyroux, T. Despeyroux, and G. Kahn. A simple applicative language: Mini-ML. In *Proc. ACM Conf. Lisp and Functional Programming*, pages 13–27, 1986.

[CF91] P. H. Cheong and L. Fribourg. Efficient integration of simplification into prolog. In J. Maluszyński and M. Wirsing, editors, *Proceedings of the 3rd Int. Symposium on Programming Language Implementation and Logic Programming, PLILP91, Passau, Germany*, pages 359–370. Springer LNCS 528, August 1991.

[Chu40] A. Church. A formulation of the simple theory of types. *J. Symbolic Logic*, 5:56–68, 1940.

[CKW89] W. Chen, M. Kifer, and D. S. Warren. HiLog: A First-Order Semantics for Higher-Order Logic Programming Constructs. In Ewing L. Lusk and Ross A. Overbeek, editors, *Proceedings of the North American Conference on Logic Programming*. MIT Press, 1989.

[CM84] W. F. Clocksin and C. S. Mellish. *Programming in Prolog*. Springer-Verlag, 2nd edition, 1984.

[Cur93] R. Curien. Second-order E-matching as a tool for automated theorem proving. In *EPIA '93*. Springer LNCS 725, 1993.

[dB72] N. G. de Bruijn. Lambda calculus notation with nameless dummies, a tool for automatic formula manipulation, with application to the Church-Rosser theorem. *Indagationes Mathematicae*, 34:381–392, 1972.

[DFH+93] G. Dowek, A. Felty, H. Herbelin, G. Huet, C. Murthy, C. Parent, C. Paulin-Mohring, and B. Werner. The Coq proof assistant user's guide version 5.8. Technical Report 154, INRIA, May 1993.

[DHK95] G. Dowek, T. Hardin, and C. Kirchner. Higher-order unification via explicit substitutions. In *Proc. 10th Annual IEEE Symposium on Logic in Computer Science*. IEEE Computer Society Press, July 1995.

[DJ90] N. Dershowitz and J.-P. Jouannaud. Rewrite systems. In Jan Van Leeuwen, editor, *Handbook of Theoretical Computer Science, Volume B: Formal Models and Semantics*, pages 243–320. Elsevier, 1990.

[DJ95] D. J. Dougherty and P. Johann. A combinatory logic approach to higher-order *E*-unification. *Theoretical Computer Science*, 139(1–2):207–242, 6 March 1995.

[DL86] D. DeGroot and G. Lindstrom. *Logic Programming, Functions, Relations, and Equations*. Prentice-Hall, 1986.

[DM79] N. Dershowitz and Z. Manna. Proving termination with multiset orderings. *Communications of the ACM*, 22(8):465–476, 1979.

[DM93] N. Dershowitz and S. Mitra. Higher-order and semantic unification. In *Proceedings of the Thirteenth Conference on Foundations of Software Technology and Theoretical Computer Science (Bombay, India)*. Springer LNCS 751, December 1993.

[DMS92] N. Dershowitz, S. Mitra, and G. Sivakumar. Decidable matching for convergent systems (preliminary version). In Deepak Kapur, editor, *11th International Conference on Automated Deduction*, Springer LNAI 607, pages 589–602, June 15–18, 1992.

[DO90] N. Dershowitz and M. Okada. A rationale for conditional equational programming. *Theoretical Computer Science*, 75(1):111–138, 1990.

[DOS88] N. Dershowitz, M. Okada, and G. Sivakumar. Canonical conditional rewrite systems. In *Proc. of the 9th International Conference on Automated Deduction*. Springer LNCS 310, 1988.

[Dow92] G. Dowek. Third order matching is decidable. In *Proceedings, Seventh Annual IEEE Symposium on Logic in Computer Science*, pages 2–10, Santa Cruz, California, 22–25 June 1992. IEEE Computer Society Press.

[Dow93] G. Dowek. Personal communication. 1993.

[dSD94] D. de Schreyeand and S. Decorte. Termination of logic programs: The never-ending story. *Journal of Logic Programming*, 19:199–260, 1994.

[DW88] M. R. Donat and L. A. Wallen. Learning and applying generalised solutions using higher order resolution. In E. Lusk and R. Overbeek, editors, *9th International Conference On Automated Deduction*, pages 41–61. Springer-Verlag, 1988.

[EM85] H. Ehrig and B. Mahr. *Fundamentals of Algebraic Specification 1*. Springer-Verlag, 1985.

[Far88] W. M. Farmer. A unification algorithm for second-order monadic terms. *Annals of Pure and Applied Logic*, 39:131–174, 1988.

[Far91] W. M. Farmer. Simple second-order languages for which unification is undecideable. *Theoretical Computer Science*, 87:25–41, 1991.

[Fay79] M. Fay. First order unification in equational theories. In *Proc. 4th Conf. on Automated Deduction*, pages 161–167. Academic Press, 1979.

[Fel92] A. Felty. A logic-programming approach to implementing higher-order term rewriting. In L.-H. Eriksson, L. Hallnäs, and P. Schroeder-Heister, editors, *Extensions of Logic Programming, Proc. 2nd Int. Worksho p*, pages 135–158. LNCS 596, 1992.

[FH86] F. Fages and G. Huet. Complete sets of unifiers and matchers in equational theories. *Theoretical Computer Science*, 43:189–200, 1986.

[FH88] A. J. Field and P. G. Harrison. *Functional Programming*. Addison-Wesley, Wokingham, 1988.

[FH91] U. Fraus and H. Hußmann. A narrowing-based theorem prover. In *Rewriting Techniques an Applications*, pages 435–436. Springer LNCS 488, April 1991.

[Fri85] L. Fribourg. SLOG: A logic programming language interpreter based on clausal superposition and rewriting. In *Symposium on Logic Programming*. IEEE Computer Society, July 1985.

[GHR92] J. C. GonzálezMoreno, M. T. HortaláGonzález, and M. RodríguezArtalejo. On the completeness of narrowing as the operational semantics of functional logic programming. In E. Börger et al., editor, *CSL'92*, Springer LNCS, San Miniato, Italy, September 1992.

[GHR97] J. C. González-Moreno, M. T. Hortalá-González, and M. Rodríguez-Artalejo. A higher order rewriting logic for functional logic programming. In *Proc. International Conference on Logic Programming*. MIT Press, 1997.

[GK96] C. Gardent and M. Kohlhase. Focus and higher–order unification. In *Proceedings of the 16th International Conference on Computational Linguistics*, Copenhagen, Denmark, 1996.

[GKL96] C. Gardent, M. Kohlhase, and N. Leusen. Corrections and higher–order unification. In *3. Konferenz zur Verarbeitung Natuerlicher Sprache*, Bielefeld, Germany, 1996.

[GLMP91] E. Giovannetti, G. Levi, C. Moiso, and C. Palamidessi. Kernel-LEAF: A logic plus functional language. *Journal of Computer and System Sciences*, 42(2):139–185, April 1991.

[GLT89] J.-Y. Girard, Y. Lafont, and P. Taylor. *Proofs and Types*. Cambridge Tracts in Theoretical Computer Science 7. Cambridge University Press, 1989.

[Gol81] W. D. Goldfarb. The undecidability of the second-order unification problem. *Theoretical Computer Science*, 13:225–230, 1981.

[Gor88] M. J. Gordon. HOL: A proof generating system for higher-order logic. In G. Birtwistle et al., editor, *VLSI Specification, Verification and Synthesis*. Kluwer Academic Press, 1988.

[Gou66] W. E. Gould. A matching procedure for ω-order logic. Scientific Report 4, Air Force Cambridge Research Laboratories, 1966.

[GTW89] J. A. Goguen, J. W. Thatcher, and E. G. Wagner. An initial algebra approach to the specification, correctness and implementation of abstract data types. In R. T. Yeh, editor, *Current trends in programming methodology*, volume 3, Data structuring, pages 80–149. Prentice-Hall, 1989.

[Hag91a] M. Hagiya. From programming-by-example to proving-by-example. In *International Conference on Theoretical Aspects of Computer Software*, pages 387–419. Springer LNCS 526, 1991.

[Hag91b] M. Hagiya. Synthesis of rewrite programs by higher-order and semantic unification. *New Generation Computing*, 8, 1991.

[Han91] M. Hanus. Efficient implementation of narrowing and rewriting. In *Proc. Int. Workshop on Processing Declarative Knowledge*, pages 344–365. Springer LNAI 567, 1991.

[Han92] M. Hanus. Improving control of logic programs by using functional logic languages. In *Proc. of the 4th International Symposium on Programming Language Implementation and Logic Programming*, pages 1–23. Springer LNCS 631, 1992.

[Han94a] M. Hanus. Combining lazy narrowing and simplification. In *Proc. 6th International Symposium on Programming Language Implementation and Logic Programming*, pages 370–384. Springer LNCS 844, 1994.

[Han94b] M. Hanus. The integration of functions into logic programming: From theory to practice. *Journal of Logic Programming*, 19&20:583–628, 1994.

[Han94c] M. Hanus. Lazy unification with simplification. In *Proc. 5th European Symposium on Programming*, pages 272–286. Springer LNCS 788, 1994.

[Han97] M. Hanus. A unified computation model for functional and logic programming. In *Proc. 24st ACM Symposium on Principles of Programming Languages (POPL'97)*, pages 80–93. ACM Press, 1997.

[Har90] M. Harao. Analogical reasoning based on higher-order unification. In S. Arikawa, S. Goto, S. Ohsuga, and T . Yokomori, editors, *Algorithmic Learning Theory*, pages 151–163. Springer-Verlag, 1990.

[HKMN95] M. Hanus, H. Kuchen, and J.J. Moreno-Navarro. Curry: A truly functional logic language. In *Proc. ILPS'95 Workshop on Visions for the Future of Logic Programming*, pages 95–107, 1995.

[HL78] G. Huet and B. Lang. Proving and applying program transformations expressed with second-order patterns. *Acta Informatica*, 11:31–55, 1978.

[HL91] G. Huet and J.-J. Lévy. Computations in orthogonal rewriting systems, I. In J.-L. Lassez and G. Plotkin, editors, *Computational Logic: Essays in Honor of Alan Robinson*, pages 395–414. MIT Press, Cambridge, MA, 1991.

[HM88] J. Hannan and D. Miller. Uses of higher-order unification for implementing program transformers. In *Fifth International Logic Programming Conference*, pages 942–959, Seattle, Washington, August 1988. MIT Press.

[Höl88] S. Hölldobler. From paramodulation to narrowing. In Robert A. Kowalski and Kenneth A. Bowen, editors, *Proceedings of the Fifth International Conference and Symposium on Logic Programming*, pages 327–342, Seatle, 1988. ALP, IEEE, The MIT Press.

[Höl89] S. Hölldobler. *Foundations of Equational Logic Programming*. Springer LNCS 353, 1989.

[HP96] M. Hanus and C. Prehofer. Higher-order narrowing with definitional trees. In *Proc. Seventh International Conference on Rewriting Techniques and Applications (RTA'96)*. Springer LNCS 1103, 1996.

[HS86] J.R. Hindley and J. P. Seldin. *Introduction to Combinators and λ-Calculus*. Cambridge University Press, 1986.

[Hsi85] J. Hsiang. Refutational theorem proving using term-rewriting systems. *Artificial Intelligence*, 25:255–300, 1985.

[Hue73] G. Huet. The undecidability of unification in third order logic. *Information and Control*, 22:257–267, 1973.

[Hue75] G. Huet. A unification algorithm for typed λ-calculus. *Theoretical Computer Science*, 1:27–57, 1975.

[Hue76] G. Huet. *Résolution d'équations dans les languages d'ordre 1,2,...ω*. PhD thesis, University Paris-7, 1976.

[Hue80] G. Huet. Confluent reductions: Abstract properties and applications to term rewriting systems. *Journal of the ACM*, 27:797–821, 1980.

[Hug86] R. J. M. Hughes. A novel representation of lists and its application to the function "reverse". *Information Processing Letters*, pages 141–144, 1986.

[Hul80] J.-M. Hullot. Canonical forms and unification. In W. Bibel and R. Kowalski, editors, *Proceedings of 5th Conference on Automated Deduction*, pages 318–334. Springer LNCS 87, 1980.

[Huß93] H. Hußmann. *Nondeterminism in Algebraic Specifications and Algebraic Programs*. Birkhäuser, 1993.

[JK86] J.-P. Jouannaud and C. Kirchner. Completion of a set of rules modulo a set of equations. *SIAM Journal of Computing*, 15(4):1155–1194, 1986.

[JK91] J.-P. Jouannaud and C. Kirchner. Solving equations in abstract algebras: A rule-based survey of unification. In J.-L. Lassez and G. Plotkin, editors, *Computational Logic: Essays in Honor of Alan Robinson*, pages 257–321. MIT Press, 1991.

[Jon92] S. Peyton Jones. Implementing lazy functional languages on stock hardware: the spineless tagless G-machine. *J. Functional Programming*, 2(2):127–202, 1992.

[JP76] D. Jensen and T. Pietrzykowski. Mechanizing ω-order type theory through unification. *Theoretical Computer Science*, 3, 1976.

[JR96] J.-P. Jouannaud and A. Rubio. A recursive path ordering for higher-order terms in η-long β-normal form. In H. Ganzinger, editor, *Proc. Seventh International Conference on Rewriting Techniques and Applications (RTA'96)*. Springer LNCS 1103, 1996.

[KA96] H. Kuchen and J. Anastasiadis. Higher Order Babel — language and implementation. In *Proceedings of Extensions of Logic Programming*. Springer LNCS 1050, 1996.

[Kah95] S. Kahrs. Towards a domain theory for termination proofs. In J. Hsiang, editor, *Rewriting Techniques and Applications*. Springer LNCS 914, 1995.

[KB70] D. E. Knuth and P. B. Bendix. Simple word problems in universal algebra. In J. Leech, editor, *Computational Problems in Abstract Algebra*, pages 263–297. Pergamon Press, 1970.

[Klo80] J. W. Klop. *Combinatory Reduction Systems*. Mathematical Centre Tracts 127. Mathematisch Centrum, Amsterdam, 1980.

[Klo92] J. W. Klop. Term rewriting systems. In Samson Abramsky, Dov M. Gabbay, and T.S.E. Maibaum, editors, *Handbook of Logic in Computer Science*, volume 2, pages 2–116. Oxford University Press, 1992.

[KOR93] J. W. Klop, V. van Oostrom, and F. van Raamsdonk. Combinatory reduction systems: Introduction and survey. *Theoretical Computer Science*, 121:279–308, 1993.

[Lau93] J. Launchbury. A natural semantics for lazy evaluation. In *Conference Record of the Twentieth Annual ACM SIGPLAN-SIGACT Symposium on Principles of Programming Languages*, pages 144–154, New York, NY, 1993. ACM.

[Llo87] J. W. Lloyd. *Foundations of Logic Programming*. Springer-Verlag, 2nd edition, 1987.

[Llo94] J. W. Lloyd. Combining functional and logic programming languages. In *Proceedings of the 1994 International Logic Programming Symposium, ILPS'94*, 1994.

[LLR93] R. Loogen, F. López-Fraguas, and M. Rodríguez-Artalejo. A demand driven strategy for lazy narrowing. In *PLILP*, Tallin, Estonia, 1993. Springer LNCS.

[Loc93] H. C. Lock. *The Implementation of Functional Logic Languages*. Oldenbourg Verlag, 1993.

[LP95] O. Lysne and J. Piris. A termination ordering for higher order rewrite systems. In J. Hsiang, editor, *Rewriting Techniques and Applications*. Springer LNCS 914, 1995.

[LPL96] H. Linnestad, C. Prehofer, and O. Lysne. Higher-order proof by consistency. In *Proc. 16th Conf. Foundations of Software Technology and Theoretical Computer Science*. Springer LNCS 1180, 1996.

[LRD95] P. Lescanne and J. Rouyer-Degli. Explicit substitutions with de Bruijn's levels. In J. Hsiang, editor, *Rewriting Techniques and Applications*. Springer LNCS 914, 1995.

[LS93] C. A. Loría-Sáenz. *A Theoretical Framework for Reasoning about Program Construction Based on Extensions of Rewrite Systems*. PhD thesis, Univ. Kaiserslautern, December 1993.

[Luc72] C. L. Lucchesi. The undecidability of the unification problem for third order languages. Technical Report CSRR 2059, University of Waterloo, Waterloo, Canada, 1972.

[Mak77] G. S. Makanin. The problem of solvability of equations in a free semigroup. *Math. USSR Sbornik*, 32:129–198, 1977.

[Mar90] L. Maranget. Optimal derivations in weak lambda-calculi and in orthogonal term rewriting systems. In *Proc. ACM Symposium on Principles of Programming Languages*, pages 255–269. ACM Press, 1990.

[MH94] A. Middeldorp and E. Hamoen. Completeness results for basic narrowing. *J. of Applicable Algebra in Engineering, Communication and Computing*, 5:213–253, 1994. Short version appeared at ALP '92.

[Mil91a] D. Miller. A logic programming language with lambda-abstraction, function variables, and simple unification. *J. Logic and Computation*, 1:497–536, 1991.

[Mil91b] D. Miller. Unification of simply typed lambda-terms as logic programming. In P.K. Furukawa, editor, *Proc. 1991 Joint Int. Conf. Logic Programming*, pages 253–281. MIT Press, 1991.

[Mil92] D. Miller. Unification under a mixed prefix. *Journal of Symbolic Computation*, 14(4):321–358, October 1992.

[MN97] R. Mayr and T. Nipkow. Higher-order rewrite systems and their confluence. *Theoretical Computer Science*, 1997.

[MNRA92] J. J. Moreno-Navarro and M. Rodriguez-Artalejo. Logic programming with functions and predicates: The language BABEL. *The Journal of Logic Programming*, 12(1, 2, 3 and 4):191–223, 1992.

[MP92a] S. Michaylov and F. Pfenning. An empirical study of the runtime behavior of higher-order logic programs. In D. Miller, editor, *Proceedings of the Workshop on the Lambda Prolog Programming Language*, Philadelphia, Pennsylvania, July 1992. University of Pennsylvania.

[MP92b] S. Michaylov and F. Pfenning. Natural semantics and some of its meta-theory in Elf. In L.-H. Eriksson, L. Hallnäs, and P. Schroeder-Heister, editors, *Extensions of Logic Programming, Proc. 2nd Int. Workshop*, pages 299–344. Springer LNCS 596, 1992.

[MP93] S. Michaylov and F. Pfenning. Higher-order logic programming as constraint logic programming. In *Position Papers for the First Workshop on Principles and Practice of Constraint Programming*, pages 221–229, Newport, Rhode Island, April 1993. Brown University.

[MPW92a] R. Milner, J. Parrow, and D. Walker. A calculus of mobile processes, part I. *Information and Computation*, 100(1):1–40, 1992.

[MPW92b] R. Milner, J. Parrow, and D. Walker. A calculus of mobile processes, part II. *Information and Computation*, 100(1):41–77, 1992.

[MRM89] A. Martelli, G. F. Rossi, and C. Moiso. Lazy unification algorithms for canonical rewrite systems. In H. Aït-Kaci and M. Nivat, editors, *Resolution of Equations in Algebraic Structures, Vol. 2, Rewriting Techniques*. Academic Press, 1989.

[MTH90] R. Milner, M. Tofte, and R. Harper. *The Definition of Standard ML*. MIT Press, 1990.

[MW94] O. Müller and F. Weber. Theory and praxis of minimal modular higher-order E-unification. In *Automated Deduction — CADE-12*. Springer LNAI 814, 1994.

[Nad87] G. Nadathur. *A Higher-Order Logic as the Basis for Logic Programming*. PhD thesis, University of Pennsylvania, Philadelphia, 1987.

[Nar89] P. Narendran. Some remarks on second order unification. Technical report, Institute of Programming and Logics, Dep. of Computer Science, State Univ. of New York at Albany, 1989.

[Nip91a] T. Nipkow. Higher-order critical pairs. In *Proc. 6th IEEE Symp. Logic in Computer Science*, 1991.

[Nip91b] T. Nipkow. Higher-order unification, polymorphism, and subsorts. In S. Kaplan and M. Okada, editors, *Proc. 2nd Int. Workshop Conditional and Typed Rewriting Systems*. Springer LNCS 516, 1991.

[Nip93a] T. Nipkow. Functional unification of higher-order patterns. In *Proc. 8th IEEE Symp. Logic in Computer Science*, pages 64–74, 1993.

[Nip93b] T. Nipkow. Functional unification of higher-order patterns. In *Proceedings, Eighth Annual IEEE Symposium on Logic in Computer Science*, Montreal, Canada, 19–23 June 1993. IEEE Computer Society Press.

[Nip93c] T. Nipkow. Orthogonal higher-order rewrite systems are confluent. In M.A. Bezem and Jan Friso Groote, editors, *Proc. Int. Conf. Typed Lambda Calculi and Applications*, pages 306–317. Springer LNCS 664, 1993.

[NJW93] G. Nadathur, B. Jayaraman, and D. S. Wilson. Implementation considerations for higher-order features in logic programming. Technical Report CS-1993-16, Duke University, 1993.

[NM88] G. Nadathur and D. Miller. An overview of λ-Prolog. In Robert A. Kowalski and Kenneth A. Bowen, editors, *Proc. 5th Int. Logic Programming Conference*, pages 810–827. MIT Press, 1988.

[NMI95] K. Nakahara, A. Middeldorp, and T. Ida. A complete narrowing calculus for higher-order functional logic programming. In *Proceedings of Seventh International Conference on Programming Language Implementation and Logic Programming 95 (PLILP'95), Springer LNCS 982*, pages 97–114, 1995.

[NP95] T. Nipkow and C. Prehofer. Type reconstruction for type classes. *J. Functional Programming*, 5(2):201–224, 1995. Short version appeared in POPL '93.

[NS94] J. Niehren and G. Smolka. A confluent calculus for higher-order concurrent constraint programming. In J.-P. Jouannaud, editor, *1st International Conference on Constraints in Computational Logics*. Springer LNCS 845, 1994.

[OMI95] S. Okui, A. Middeldorp, and T. Ida. Lazy narrowing: Strong completeness and eager variable elimination. In *Proceedings of the 20th Colloquium on Trees in Algebra and Programming*. Springer LNCS 915, 1995.

[Oos94] V. van Oostrom. *Confluence for Abstract and Higher-Order Rewriting*. PhD thesis, Vrije Universiteit, 1994. Amsterdam.

[Oos96] V. van Oostrom. Higher-order families. In H. Ganzinger, editor, *Proc. Seventh International Conference on Rewriting Techniques and Applications (RTA'96)*. Springer LNCS 1103, 1996.

[OR94a] V. van Oostrom and F. van Raamsdonk. Comparing combinatory reduction systems and higher-order rewrite systems. In J. Heering, K. Meinke, B. Möller, and T. Nipkow, editors, *Higher-Order Algebra, Logic and Term Rewriting*, pages 276–304. Springer LNCS 816, 1994.

[OR94b] V. van Oostrom and F. van Raamsdonk. Weak orthogonality implies confluence: the higher-order case. In A. Nerode, editor, *Logical Foundations of Computer Science*, pages 379–392. Springer LNCS 813, 1994.

[Pad95] V. Padovani. On equivalence classes of interpolation equations. In M. Dezani-Ciancaglini and G. Plotkin, editors, *Typed Lambda Calculi and Applications*, pages 335–349. Springer LNCS 902, 1995.

[Pau86] L. C. Paulson. Natural deduction proof as higher-order resolution. *Journal of Logic Programming*, 3:237–258, 1986.

[Pau90] L. C. Paulson. Isabelle: The next 700 theorem provers. In P. Odifreddi, editor, *Logic and Computer Science*, pages 361–385. Academic Press, 1990.

[Pau91] L. C. Paulson. *ML for the Working Programmer*. Cambridge University Press, 1991.

[Pau94] L. C. Paulson. *Isabelle: A Generic Theorem Prover*. Springer LNCS 828, 1994.

[PE88] F. Pfenning and C. Elliott. Higher-order abstract syntax. In *Proc. SIGPLAN '88 Symp. Programming Language Design and Implementation*, pages 199–208. ACM Press, 1988.

[Pfe88] F. Pfenning. Partial polymorphic type inference and higher-order unification. In *ACM Conference on Lisp and Functional Programming*, pages 153–163, Snowbird, Utah, July 1988. ACM-Press.

[Pfe91] F. Pfenning. Logic programming in the LF logical framework. In G. Huet and G. Plotkin, editors, *Logical Frameworks*. Cambridge University Press, 1991.

[PHA+96] J. Peterson[editor], K. Hammond[editor], L. Augustsson, B. Boutel, W. Burton, J. Fasel, A. Gordon, J. Hughes, P. Hudak,

T. Johnsson, M. Jones, S. Peyton Jones, A. Reid, and P. Wadler. Haskell 1.3, A non-strict, purely functional language. Report YALEU / DCS / RR-1106, Department of Computer Science, Yale University, May 1996.

[Pie73] T. Pietrzykowski. A complete mechanization of second-order type theory. *J. of ACM*, 20:333–364, 1973.

[Plo75] G. Plotkin. Call-by-name, call-by-value and the λ-calculus. *Theoretical Computer Science*, 1:125–159, 1975.

[PM90] R. Pareschi and D. Miller. Extending definite clause grammars with scoping constructs. In David H. D. Warren and Peter Szeredi, editors, *1990 International Conference in Logic Programming*, pages 373–389. MIT Press, June 1990.

[Pol94] J. van de Pol. Termination proofs for higher-order rewrite systems. In J. Heering, K. Meinke, B. Möller, and T. Nipkow, editors, *Higher-Order Algebra, Logic and Term Rewriting*, volume 816 of *Lect. Notes in Comp. Sci.*, pages 305–325. Springer-Verlag, 1994.

[Pre94a] C. Prehofer. Decidable higher-order unification problems. In *Automated Deduction — CADE-12*. Springer LNAI 814, 1994.

[Pre94b] C. Prehofer. Higher-order narrowing. In *Proc. Ninth Annual IEEE Symposium on Logic in Computer Science*. IEEE Computer Society Press, July 1994.

[Pre94c] C. Prehofer. On modularity in term rewriting and narrowing. In J.-P. Jouannaud, editor, *1st International Conference on Constraints in Computational Logics*, München, Germany, 7–9 September 1994. Springer LNCS 845.

[Pre95a] C. Prehofer. A Call-by-Need Strategy for Higher-Order Functional-Logic Programming. In J. Lloyd, editor, *Logic Programming. Proc. of the 1995 International Symposium*, pages 147–161. MIT Press, 1995.

[Pre95b] C. Prehofer. Higher-order narrowing with convergent systems. In *4th Int. Conf. Algebraic Methodology and Software Technology, AMAST '95*. Springer LNCS 936, July 1995.

[Pre95c] C. Prehofer. *Solving Higher-order Equations: From Logic to Programming*. PhD thesis, TU München, 1995. Also appeared as Technical Report I9508.

[Pre97] C. Prehofer. Higher-order functional-logic programming: A systematic development (invited talk). In *Proc. of the Second Fuji International Workshop on Functional and Logic Programming*. World Scientific, 1997.

[PS95] J. van de Pol and H. Schwichtenberg. Strict functionals for termination proofs. In M. Dezani-Ciancaglini and G. Plotkin, editors, *Typed Lambda Calculi and Applications*, pages 350–364. Springer LNCS 902, 1995.

[Qia93] Z. Qian. Linear unification of higher-order patterns. In M.-C. Gaudel and J.-P. Jouannaud, editors, *Proceedings of the Colloquium on Trees in Algebra and Programming*, pages 391–405, Orsay, France, April 1993. Springer-Verlag LNCS 668.

[Qia94] Z. Qian. Higher-order equational logic programming. In *Proc. 21st ACM Symposium on Principles of Programming Languages*, Portland, 1994. ACM Press.

[QW94] Z. Qian and K. Wang. Modular AC unification of higher-order patterns. In J.-P. Jouannaud, editor, *1st International Conference on Constraints in Computational Logics*, München, Germany, 1994. Springer LNCS 845.

[Raa93] F. van Raamsdonk. Confluence and superdevelopments. In *Rewriting Techniques an Applications*, pages 168–182. LNCS 690, June 1993.

[Red85] U. S. Reddy. Narrowing as the operational semantics of functional languages. In *Symposium on Logic Programming*. IEEE Computer Society, July 1985.

[Red86] U. S. Reddy. On the relationship between logic and functional languages. In D. DeGroot and G. Lindstrom, editors, *Logic Programming: Functions, Relations, and Equations*, pages 3–36. Prentice-Hall, Englewood Cliffs, NJ, 1986.

[Red94] U. S. Reddy. Higher-order aspects of logic programming. In *ICLP'94*. MIT Press, Cambridge, MA, 1994.

[RW69] G. A. Robinson and L. T. Wos. Paramodulation and theorem proving in first order theories. In B. Meltzer and D. Michie, editors, *Machine Intelligence 4*, pages 133–150. American Elsevier, 1969.

[SG89] W. Snyder and J. Gallier. Higher-order unification revisited: Complete sets of transformations. *J. Symbolic Computation*, 8:101–140, 1989.

[She90] Y.-H. Sheng. HIFUNLOG: Logic programming with higher-order relational functions. In David H.D. Warren and Peter Szeredi, editors, *Logic Programming*. MIT Press, 1990.

[SJ92] F. S. K. Silbermann and B. Jayaraman. A domain-theoretic approach to functional and logic programming. *Journal of Functional Programming*, 2(3):273–321, 1992.

[Sla74] J. R. Slagle. Automated theorem-proving for theories with simplifiers, commutativity, and associativity. *Journal of the ACM*, 1974.

[Smo86] G. Smolka. Fresh: A higher-order language based on unification. In Doug DeGroot and Gary Lindstrom, editors, *Logic Programming Functions, Relations, and Equations*, pages 469–525. Prentice-Hall, 1986.

[Smo94] G. Smolka. The Definition of Kernel Oz. Technical Report DFKI Research Report RR-94-23, Deutsches Forschungsinstitut für Künstliche Intelligenz, Stuhlsatzenhausweg 3, D-66123 Saarbrücken, Germany, 1994.

[Sny90] W. Snyder. Higher order E-unification. In M. E. Stickel, editor, *10th International Conference on Automated Deduction*, pages 573–587, Berlin, Heidelberg, 1990. Springer LNAI 449.

[Sny91] W. Snyder. *A Proof Theory for General Unification*. Birkhäuser, Boston, 1991.

[SS86] L. Sterling and E. Shapiro. *The Art of Prolog: Advanced Programming Techniques*. MIT Press, 1986.

[Ste90] G. L. Steele. *Common LISP: The Language (Second Edition)*. Digital Press, Burlington, MA, 1990.

[Tho90] W. Thomas. Automata on infinite objects. In Jan Van Leeuwen, editor, *Handbook of Theoretical Computer Science Volume B: Formal Models and Semantics*, pages 134–191. Elsevier, 1990.

[Tur86] D. Turner. An overview of Miranda. *Sigplan Notices*, 21(12):158–160, 1986.

[Wad71] C. P. Wadsworth. *Semantics and Pragmatics of the Lambda Calculus*. Phd thesis, University of Oxford, Oxford, September 1971.

[Wad90] P. Wadler. Deforestation: transforming programs to eliminate trees. *Theoretical Computer Science*, 73(2):231–248, June 1990.

[War82] D. H. D. Warren. Higher-order extensions to prolog: Are they needed? In D. Michie, editor, *Machine Intelligence*, pages 441–454. Edinburgh University Press, 1982.

[Wir90] M. Wirsing. Algebraic Specification. In J. van Leeuwen, editor, *Handbook of Theoretical Computer Science.*, chapter 13, pages 675–788. North-Holland, Amsterdam, 1990.

[Wol93] D. A. Wolfram. *The Clausal Theory of Types*. Cambridge Tracts in Theoretical Computer Science 21. Cambridge University Press, 1993.

Index

$(\longrightarrow_1, \longrightarrow_2)_{lex}$, 26
$<_{sub}$, 31
$=^?$, 33
$=_E$, 33
$=_W$, 32
$=_s$, 119
$=_{E,W}$, 33
Dom, 32
E-unification, 33
R-normal form, 48
R-normalized, 48
$[t \mid R]$, 27
$\mathcal{BV}$, 27, 31
$\mathcal{FV}$, 27, 32
Im, 32
Irr, 96
$\mathcal{OBV}$, 68
$\mathcal{R}ng$, 32
$\overset{*}{\Rightarrow}$, 39, 82
α-conversion, 28
β-conversion, 12, 28
β-normal form, 29
β-redex, 29
βη-normal form
  long, 29
$\downarrow_R$, 26, 48
$\rightsquigarrow$, 126
$\overset{p}{\rightsquigarrow}$, 122
ε, 30
η-conversion, 12, 28
η-expanded form, 29
η-expansion, 29
η-normal form, 29

$\rightarrow^?$, 81, 119
$\overset{?}{\leftrightarrow}$, 81
λ-Prolog, 2, 15, 158
λ-calculus
  conversions, 28
  weak, 158
λ-term, *see* term
$t\updownarrow^{\eta}_{\beta}$, 29
$\leq_W$, 32
$\leq_{E,W}$, 33
$\leq_W$, 33
$|\, t \,|$, 29
$|_W$, 32
$|_p$, 31
$\overline{x_k}$-lifter, 47
$\longrightarrow$, 46
$\longrightarrow^R_{sub}$, 52
$\longrightarrow_{\neq\varepsilon}$, 48
$\rightarrow$ (type constructor), 27
$\tau_{F,i}$, 58
$t[s]_p$, 31

abstract reduction, 25
abstract syntax
  higher-order, 151
abstraction, 27
alternating bit protocol, 146
application, 27
Application, rule, 49
atom, 27
authorization, 143

BABEL, 157
backtracking, 114

CHRS, 112
CLN, 114
communication
  asynchronous, 147
confluence, 26
  ground, 51
  local, 26
congruence, 49
constant, 27
Constraint, rule
  Failure, 95
  Solving, 131
constructor, 48
Constructor, rule
  Clash, 87
  Decomposition, 87
  Imitation, 87
  Occurs Check, 87
convergent, 52
Conversion, rule, 49
critical pair, 51
cycle free, 97

dag-solved form, 101
Decomposition, rule, 38, 45, 83
  Constructor, 87
Deletion, rule, 38, 45, 83
deterministic, 40
difference list, 145
differentiation, 138
distributed systems, 146

EL, rule, 58
Elf, 158
Eliminate, rule, 58
elimination problems, 55
Elimination, rule, 38, 45, 83, 95
encryption, 143
equality
  join, 112
  normal, 112
  strict, 119
equation
  flex-flex, 39
  flex-rigid, 37
equational theory, 33
equivalence relation, 25
evaluation
  call-by-name, 48
  eager, 48
  forcing, 119
expansion, 29
extensionality, 12, 29

finitary, 17
first-order
  quasi, 73
Flatten, rule, 131
flattening, 46, 130
Flex-Flex Diff, rule, 45
Flex-Flex Same, rule, 45
function
  monotonic, 52

GHRS, 46
goal, 81, 119
  connection, 97
  oriented, 120
  parallel, 97
  selection, 40, 86
  solution, 120
goal selection, 43

Haskell, 8, 97, 136
head, 29
HRS, 47
  left-linear, 48
  normal conditional, 113
  orthogonal, 51
  pattern, 48

idempotent, 32
imitation binding, 38
Imitation, rule, 38, 83

Constructor, 87
with Constraints, 95
Imitation/Projection, rule, 45
instance, 32
Isabelle, 15, 75
isolated, 29

joinable, 26

Lazy Narrowing, rule
at Variable, 83
General, 84, 95
Normalizing, 91
with Constraints, 95
with Decomposition, 83
let-construct, 22
definition, 137, 156
for pairs, 137
lexicographic ordering, 26
lhs, 62
lifting, 47
linear, 29
linear second-order system, 70
LISP, 8
list, 27
LN, 83
LNC, 95
LNN, 88
logic programming, 10
higher-order, 158
long $\beta\eta$-normal form, 29

matching, 16
MCSU, 33
more general, 33
multiset, 26
extension, 26
smaller, 26

naming conventions, 27
narrowing
completeness, 119
completeness w.r.t. solutions, 119
lazy, 120
plain, 10, 122
narrowing step, 122
pattern, 126
NC, 131
NCHRS, 113
NCLN, 117
NLN, 91
non-determinism, 146
normal form, 25
Nuprl, 15

Occurs Check, rule
Constructor, 87
ordering
compatible, 25
partial, 25
strict, 25
preserving, 100
terminating, 25
termination, 52
decreasing, 117
total, 25
overlap, 50

parsing, 141
partial binding, 39
path, 30
rigid, 31
pattern, 16, 34
relaxed, 34
Pattern Narrow, rule, 131
polymorphism, 160
position, 31
independent, 31
root, 31
postfix, 30
Proceed, rule, 58
program transformation, 149
projection binding, 38
Projection, rule, 38, 83

second-order, 39
Prolog, 114, 136, 138, 153, 154
prophecy, 141
protocol, 146
PT, 38
PU, 45

reduction
abstract, 25
conditional
length, 113
Reflexivity, rule, 49
rewrite rule, 46
$\overline{x_k}$-lifted, 47
base type, 48
conditional, 112
extra variables, 22, 112
fully extended, 108
left-linear, 48
normal conditional, 21, 113
pattern, 47
rewrite step, 46
conditional, 113
innermost, 48
outermost, 48
rhs, 62
right isolated, 97
rigid, 29
root position, 31
rule, *see* rewrite rule

second-order, 28
weakly, 28
semantics
denotational, 49, 119
equational, 119
sequence, 30
empty, 30
prefix, 30
Simple System, 97
simplified, 101
simplification, 91
SML, 8, 97, 136
Solve, rule, 131
substitution, 32
approximates, 42
composition, 32
equality, 32
free variables, 32
ground, 32
idempotent, 32
more general, 32
normal form, 32
parameter eliminating, 58
restriction, 32
size increasing, 44
well-formed, 32
subterm, 31
modulo binders, 31
symbol
constructor, 48
defined, 48
Symmetry, rule, 49

term, 27
flexible, 29
fully extended, 108
ground, 31
order, 28
simply typed, 28
size, 29
termination, 25, 52, 117
theorem proving, 76
TPS, 15
Transitivity, rule, 49
type, 27
base, 27
constructor, 27
judgment, 28
order, 28
type inference, 151

unification, 33
associative, 77

equational, 33
finitary, 33
higher-order, 37
infinitary, 16, 33
lazy, 120
nullary, 16, 33
pattern, 43
pre-, 16
unitary, 16, 33
unifier, 33
pre-, 39

variable, 27
bound, 27
convention, 30
free, 17, 27
intermediate, 104
logic, 17
loose bound, 30
of interest, 104
outside bound, 30

Progress in Theoretical Computer Science

Editor
Ronald V. Book
Department of Mathematics
University of California
Santa Barbara, CA 93106

Editorial Board
Erwin Engeler
Mathematik
ETH Zentrum
CH-8092 Zurich, Switzerland

Robin Milner
Department of Computer Science
Cambridge University
Cambridge CB2 1SB, England

Jean-Pierre Jouannaud
Laboratoire de Recherche
en Informatique Bât. 490
Université de Paris-Sud
Centre d'Orsay
91405 Orsay Cedex, France

Martin Wirsing
Universität Passau
Fakultät für Mathematik
und Informatik
Postfach 2540
D-8390 Passau, Germany

Progress in Theoretical Computer Science is a series that focuses on the theoretical aspects of computer science and on the logical and mathematical foundations of computer science, as well as the applications of computer theory. It addresses itself to research workers and graduate students in computer and information science departments and research laboratories, as well as to departments of mathematics and electrical engineering where an interest in computer theory is found.

The series publishes research monographs, graduate texts, and polished lectures from seminars and lecture series. We encourage preparation of manuscripts in some form of TeX for delivery in camera-ready copy, which leads to rapid publication, or in electronic form for interfacing with laser printers or typesetters.

Proposals should be sent directly to the Editor, any member of the Editorial Board, or to: Birkhäuser Boston, 675 Massachusetts Ave., Cambridge, MA 02139. The Series includes:

1. Leo Bachmair, *Canonical Equational Proofs*
2. Howard Karloff, *Linear Programming*
3. Ker-I Ko, *Complexity Theory of Real Functions*
4. Guo-Qiang Zhang, *Logic of Domains*
5. Thomas Streicher, *Semantics of Type Theory: Correctness, Completeness and Independence Results*
6. Julian Charles Bradfield, *Verifying Temporal Properties of Systems*
7. Alistair Sinclair, *Algorithms for Random Generation and Counting*
8. Heinrich Hussmann, *Nondeterminism in Algebraic Specifications and Algebraic Programs*
9. Pierre-Louis Curien, *Categorical Combinators, Sequential Algorithms and Functional Programming*
10. J. Köbler, U. Schöning, and J. Torán, *The Graph Isomorphism Problem: Its Structural Complexity*
11. Howard Straubing, *Finite Automata, Formal Logic, and Circuit Complexity*
12. Dario Bini and Victor Pan, *Polynomial and Matrix Computations, Volume 1 Fundamental Algorithms*

13. James S. Royer and John Case, *Subrecursive Programming Systems: Complexity & Succinctness*
14. Roberto Di Cosmo, *Isomorphisms of Types*
15. Erwin Engeler et al., *The Combinatory Programme*
16. Peter W. O'Hearn and Robert D. Tennent, *Algol-like Languages, Vol. I & II*
17. Giuseppe Castagna, *Object-Oriented Programming: A Unified Foundation*
18. Franck van Breugel, *Comparative Metric Semantics of Programming Languages: Nondeterminism and Recursion*
19. Christian Prehofer, *Solving Higher-Order Equations: From Logic to Programming*